Lecture Notes in Mathematics

1554

Editors:
A. Dold, Heidelberg
B. Eckmann, Zürich
F. Takens, Groningen

A.K. Lenstra   H.W. Lenstra, Jr.   (Eds.)

The development of the number field sieve

Springer-Verlag

Berlin Heidelberg New York
London Paris Tokyo
Hong Kong Barcelona
Budapest

Editors

Arjen K. Lenstra
Room MRE-2Q334
Bellcore
445 South Street
Morristown, NJ 07960, USA
E-mail: lenstra@bellcore.com

Hendrik W. Lenstra, Jr.
Department of Mathematics
University of California
Berkeley, CA 94720, USA
E-mail: hwl@math.berkeley.edu

Mathematics Subject Classification (1991): 11Y05, 11Y40

ISBN 3-540-57013-6 Springer-Verlag Berlin Heidelberg New York
ISBN 0-387-57013-6 Springer-Verlag New York Berlin Heidelberg

© Springer-Verlag Berlin Heidelberg 1993
Printed in Germany

46/3140-543210 - Printed on acid-free paper

PREFACE

The number field sieve is an algorithm for factoring integers that John Pollard proposed in 1988. This volume contains six papers on the number field sieve. They are preceded by an annotated bibliography, to which we refer for a brief description of the contents of each individual paper.

We assume the reader to be familiar with the basic techniques that underlie modern integer factoring methods. An introduction to these techniques is given in:

> A. K. Lenstra, H. W. Lenstra, Jr., M. S. Manasse, J. M. Pollard, *The factorization of the ninth Fermat number*, Math. Comp. **61** (1993), to appear.

The same paper includes a discussion of tools from algebraic number theory that the number field sieve depends on. Comprehensive accounts of older algorithms for factoring integers and related problems, with extensive bibliographies, can be found in:

> A. K. Lenstra, H. W. Lenstra, Jr., *Algorithms in number theory*, Chapter 12 in: J. van Leeuwen (ed.), *Handbook of theoretical computer science*, Volume A, *Algorithms and complexity*, Elsevier, Amsterdam, 1990,

> C. Pomerance (ed.), *Cryptology and computational number theory*, Proc. Sympos. Appl. Math. **42**, Amer. Math. Soc., Providence, 1990.

The developments leading up to the number field sieve are sketched on pp. 11–13 below. The annotated bibliography (pp. 1–3) tells, implicitly, the recent history of the number field sieve itself.

We express our gratitude to the authors of the papers for contributing their work to this volume. In particular we wish to thank Carl Pomerance for conceiving the idea of a combined publication and for advising us in all stages of its execution. In addition, we thank Henri Cohen, Dan Gordon, Andrew Odlyzko, Jonathan Pila, and Tom Trotter for their assistance.

We gratefully acknowledge the use of the $\mathcal{AMS}$-TEX typesetting package.

The editors

CONTENTS

THE NUMBER FIELD SIEVE:
AN ANNOTATED BIBLIOGRAPHY

H. W. LENSTRA, JR.

In the present bibliography I list, in approximately chronological order, all literature that is directly related to the number field sieve.

1. J. M. Pollard, *Factoring with cubic integers*, this volume, pp. 4–10; manuscript, 6 pages, August 1988.

Pollard describes a new method for factoring integers of a special form, and he illustrates it by means of the factorization of the seventh Fermat number $F_7 = 2^{128} + 1$. He uses the ring of integers $\mathbf{Z}[\sqrt[3]{2}]$ of the number field $\mathbf{Q}(\sqrt[3]{2})$. No sieving in the number field takes place, so the name *number field sieve* is less appropriate for this early version of the method than for its descendants.

The manuscript was enclosed with a letter to A. M. Odlyzko, dated 31 August 1988, with copies to R. P. Brent, J. Brillhart, H. W. Lenstra, C. P. Schnorr, and H. Suyama. In this letter, Pollard speculated: "If F_9 is still unfactored, then it might be a candidate for this kind of method eventually?" The answer is in [7].

2. J. M. Pollard, *Factoring with cubic integers* (2), unpublished manuscript, 3 pages, December 1988.

This forms a footnote to the previous paper. It reports on the factorization of $2^{144} - 3$.

3. A. K. Lenstra, H. W. Lenstra, Jr., M. S. Manasse, J. M. Pollard, *The number field sieve*, this volume, pp. 11–42; extended abstract: Proc. 22nd Annual ACM Symp. on Theory of Computing (STOC), Baltimore, May 14–16, 1990, 564–572.

The authors describe the first large-scale implementation of Pollard's new method, with several improvements. The extended abstract contains a rough outline of a heuristic complexity analysis, which indicates that the method is, for the numbers that it applies to, faster than all other known factoring methods. The last section of the extended abstract discusses an idea of Buhler and Pomerance for extending the number field sieve to general integers, and it sketches a solution to a problem that this extension gives rise to. The final version of the paper addresses the same issues in a more detailed manner, and it mentions developments that took place since the extended abstract was written. (*Note:* the

The author thanks Johannes Buchmann, Don Coppersmith, Dan Gordon, and John Pollard for their help. He was supported by NSF under Grant No. DMS–9002939 and by NSA/MSP under Grant No. MDA90–H–4043.

terminology *pf/fp* used in the extended abstract has been switched in all later papers, including the final version.)

4.   D. M. Gordon, *Discrete logarithms in $GF(p)$ using the number field sieve*, SIAM J. Discrete Math. **6** (1993), 124–138; prepublication: 15 pages, April 27, 1990.

It is shown that the ideas underlying the number field sieve apply, in theory at least, also to the discrete logarithm problem.

5.   L. M. Adleman, *Factoring numbers using singular integers*, Proc. 23rd Annual ACM Symp. on Theory of Computing (STOC), New Orleans, May 6–8, 1991, 64–71; prepublication: TR-20, Department of Computer Science, University of Southern California, 8 pages, September 4, 1990.

Adleman suggests the use of quadratic characters in order to recognize squares in the number field. This provides an alternative solution to the problem that the idea of Buhler and Pomerance gives rise to, and it improves the conjectural run time estimate for the number field sieve as it applies to integers that are not of a special form.

6.   D. Coppersmith, *Modifications to the number field sieve*, J. Cryptology, to appear; prepublication: IBM Research Report #RC 16264, Yorktown Heights, New York, 16 pages, November 1990.

The combined use of several number fields leads to an improvement of the conjectural run time estimate of the number field sieve.

7.   A. K. Lenstra, H. W. Lenstra, Jr., M. S. Manasse, J. M. Pollard, *The factorization of the ninth Fermat number*, Math. Comp. **61** (1993), to appear.

The ninth Fermat number $F_9 = 2^{512} + 1$ was factored in 1990 by means of the number field sieve. The paper discusses several aspects of this factorization. It can be read as an introduction to the number field sieve.

8.   J. M. Pollard, *The lattice sieve*, this volume, pp. 43–49; manuscript, 7 pages, September 1991.

Pollard advocates the use of a two-dimensional sieve in order to speed up the sieving process. It is not yet clear whether the idea leads to a practical improvement; see also [12].

9.   O. Schirokauer, *On pro-finite groups and on discrete logarithms*, Ph. D. thesis, University of California, Berkeley, 68 pages, May 1992.

In the second chapter (46 pp.) of his thesis, Schirokauer considers the application of the number field sieve to the discrete logarithm problem. Through the use of *l*-adic logarithms he achieves a better conjectural run time than in [4] (see the introduction to [3]). Practical issues are not considered.

10.   J. P. Buhler, H. W. Lenstra, Jr., C. Pomerance, *Factoring integers with the number field sieve*, this volume, pp. 50–94.

This paper describes the number field sieve as it applies to integers that are not necessarily of a special form. The description incorporates Adleman's idea [5]. An elaborate complexity analysis is given, and several possible practical improvements are discussed.

11.   J.-M. Couveignes, *Computing a square root for the number field sieve*, this volume, pp. 95–102.

One step in Adleman's version of the number field sieve [5; 10] involves a computation with exceedingly large numbers. Couveignes develops a method for avoiding this.

12.   D. J. Bernstein, A. K. Lenstra, *A general number field sieve implementation*, this volume, pp. 103–126.

The title explains itself. The implementation applies to "general" integers, but almost all examples given are integers of a special form. It remains to be decided whether the number field sieve will eventually be the method of choice for large integers.

13.   D. M. Gordon, *Designing and detecting trapdoors for discrete log cryptosystems*, Advances in cryptology, Crypto '92, to appear.

Gordon discusses the applicability of the number field sieve to the construction of trapdoors in cryptological systems that are based on the discrete logarithm problem.

14.   J. Buchmann, J. Loho, J. Zayer, *An implementation of the general number field sieve*, extended abstract, Fachbereich Informatik, Universität des Saarlandes, 7 pages, May 1993.

This report describes the authors' practical experience with their first implementation. They factor three "general" integers of 29, 40, and 49 digits.

DEPARTMENT OF MATHEMATICS, UNIVERSITY OF CALIFORNIA, BERKELEY, CA 94720, U.S.A.
E-mail address: hwl@math.berkeley.edu

FACTORING WITH CUBIC INTEGERS

J. M. POLLARD

SUMMARY. We describe an experimental factoring method for numbers of form $x^3 + k$; at present we have used only $k = 2$. The method is the cubic version of the idea given by Coppersmith, Odlyzko and Schroeppel (Algorithmica 1 (1986), 1–15), in their section 'Gaussian integers'. We look for pairs of small coprime integers a and b such that:

i.   the integer $a + bx$ is smooth,

ii.   the algebraic integer $a + bz$ is smooth, where $z^3 = -k$. This is the same as asking that its norm, the integer $a^3 - kb^3$ shall be smooth (at least, it is when $k = 2$).

We used the method to repeat the factorisation of F_7 on an 8-bit computer ($2F_7 = x^3 + 2$, where $x = 2^{43}$).

INTRODUCTION

We consider the case $k = 2$ throughout. We denote by $\mathbf{Z}$ the set of rational integers (ordinary integers) and by S the set of algebraic integers:

$$[a, b, c] = a + bz + cz^2, \qquad (a, b, c \text{ in } \mathbf{Z}).$$

These constitute the algebraic integers of the field generated by z, and possess the property of unique factorisation (neither statement true for general k, see e. g. [2]). According to [1], such methods are still possible when unique factorisation fails.

We also write:

$$\{a, b, c\} = a + bx + cx^2.$$

When ii. holds, we have some factorisation:

$$[a, b, 0] = [d, e, f] \cdot \ldots,$$

into units and primes of S (defined shortly). Then also:

$$\{a, b, 0\} \equiv \{d, e, f\} \cdot \ldots \pmod{n}.$$

But by i. we have also a factorisation:

$$\{a, b, 0\} = p \cdot q \cdot \ldots,$$

into small primes of $\mathbf{Z}$. So we have a congruence (mod n) involving rational integers from two small sets. From a sufficient number of such congruences, we obtain some equations:

$$X^2 \equiv Y^2 \pmod{n},$$

and hopefully the factorisation of n.

1991 *Mathematics Subject Classification*. Primary 11Y05, 11Y40.

Key words and phrases. Factoring algorithm, algebraic number fields.

PROPERTIES OF THE SET S

The norm of a member $[a, b, c]$ of S is the rational integer:

$$N(a, b, c) = a^3 - 2b^3 + 4c^3 + 6abc.$$

This is a multiplicative function, i.e. the equation

$$[a, b, c] = [d, e, f] \cdot [g, h, i] \tag{1}$$

implies:

$$N(a, b, c) = N(d, e, f) \cdot N(g, h, i).$$

Given an equation (1), we say that $[d, e, f]$ divides $[a, b, c]$.

The norm can be zero only when $a = b = c = 0$. Numbers with norm $+1$ or -1 are called *units*. There are an infinity of units, namely all the numbers:

$$\pm U^i \quad (i = 0, \pm 1, \pm 2, \dots),$$

where $U = [1, 1, 0]$. We give a table of the small powers of U:

i	U^i	U^{-i}
0	$[\ 1, 0, 0]$	$[\quad 1, \quad 0, \quad 0]$
1	$[\ 1, 1, 0]$	$[\ -1, \quad 1, \ -1]$
2	$[\ 1, 2, 1]$	$[\quad 5, \ -4, \quad 3]$
3	$[-1, 3, 3]$	$[-19, \quad 15, -12]$
4	$[-7, 2, 6]$	$[\ 73, -58, \quad 46]$

A unit divides any integer. If $[d, e, f]$ in (1) is a unit, then the other two numbers are termed *associates*; clearly this means that:

$$N(a, b, c) = \pm N(g, h, i),$$

but the converse statement is false as we shall see.

A number $[a, b, c]$ is termed *prime* if any factorisation (1) contains a unit (and an associate). A number of norm $\pm p$ (p prime) is certainly a prime; but not all primes are of this form.

A rational prime p need not be a prime of S. We have $N(p, 0, 0) = p^3$, so perhaps p can have prime factors of norm $\pm p$ or $\pm p^2$. Indeed it can. There are four cases (see [2, p. 186]):

1. The primes $p = 2$ and 3. These factor as a unit and the cube of a prime of norm p:

$$2 = \quad -1 \quad \cdot [\ 0, 1, 0]^3,$$
$$3 = [1, 1, 0] \cdot [-1, 1, 0]^3.$$

2. Primes p of form $6m + 1$, with -2 a cubic residue (mod p):

$$p = 31, \ 43, \ 109, \ 127, \ 157, \ \dots.$$

There are three nonassociated factors of norm p. For example:

$$31 = [5, -4, 3] \cdot [-1, -2, 1] \cdot [-9, -6, 1] \cdot [3, 0, 1].$$

(The first factor on the right is a unit.)

3. Primes of form $p = 6m + 5$:

$$p = 5, 11, 17, 23, 29, \ldots.$$

There is one factor of norm p and one of norm p^2. For example:

$$5 = [1, 0, 1] \cdot [1, -2, -1].$$

4. Primes p of form $6m + 1$, with -2 a cubic nonresidue (mod p):

$$p = 7, 13, 19, 37, \ldots.$$

There is no factorisation: p is a prime of norm p^3.

OPERATIONS ON THE INTEGERS OF S

It is easy to add, subtract and multiply numbers $[a, b, c]$; when multiplying, we use $z^3 = -2$, to remove the cube and fourth power terms in z. As for division, we do not need a Euclidean algorithm (does it exist?), but only to test whether:

$$[d, e, f] | [a, b, c]? \tag{2}$$

and, if so, to find the quotient, $[g, h, i]$ in (1).

We shall use:
$$N(a, b, c) = [a, b, c] \cdot C(a, b, c),$$

where
$$C(a, b, c) = [a^2 + 2bc, -ab - 2c^2, b^2 - ac].$$

This can be easily verified; in fact C is the product of the conjugates of $[a, b, c]$, obtained by replacing z by the other two cube roots of -2. To test (2) we multiply both sides by $C(d, e, f)$; this gives the question

$$N(d, e, f) | [A, B, C]?$$

This holds when each of A, B and C is divisible by the integer $N = N(d, e, f)$: if so the required quotient is $[A/N, B/N, C/N]$.

FACTORISATION OF F_7

An obvious test case for our method is $F_7 = 2^{128} + 1$, first factored by Brillhart and Morrison in 1970 [3]. We have:

$$2F_7 = x^3 + 2, \qquad \text{where } x = 2^{43}.$$

We describe the method in detail for this number.

Step 1. Compute the factor base.

The first part of the factor base, FB_1, consists of the first 500 primes:

$$2, 3, 5, \ldots, 3571.$$

It is also convenient to compute $x \pmod{p}$ for each such prime.

The second part FB_2 consists of primes of S arising from the factorisations of the rational primes p of FB_1. Only primes of norm $\pm p$ are used. Those of norm p^2 in case 3 cannot divide numbers $[a, b, 0]$, $\gcd(a, b) = 1$, and are not needed.

Cases	Times	Primes
1	2	2
2	81	243
3	252	252
4	165	0
Totals	500	497

We also included in FB_2 three units: -1, U and $1/U$, making 500 members in all, like FB_1 (a coincidence!). The choice of FB_2 was dictated by convenience, and is larger than necessary. We have at once 81 equations connecting FB_1 and FB_2, and one more involving the units.

Step 2. Run the sieve.

We want to find numbers $a + bx$ composed of the primes of FB_1, except perhaps for one larger prime. Our program is like that for the Quadratic Sieve of Pomerance [4], but simpler; b is held constant while a varies over an interval of width up to 12,000. Only coprime pairs (a, b) are saved.

$$\text{Range for } b = 1 \ldots 2000$$
$$\text{Range for } a = -4800 \ldots 4800$$
$$\text{Limit for large prime} = 10,000$$
$$\text{Integers sieved} = 1.92(7)$$
$$\text{Successes} = 40,762$$

Note:

1. My sieve represents an integer m by the nearest integer to $2\log_2(m)$. This means that the limit on the large prime is very rough.

2. We do not know, at this stage, how many of these successes involve a large prime.

Step 2A. Look for smooth values of the norm.

For each pair, we compute the norm:

$$N = N(a, b, 0) = a^3 - 2b^3.$$

We factor by trial division, using only the norms of primes of FB_2 (there are 335 distinct values). When N factors completely over these primes (no 'large' prime allowed here), compute the large prime q, if any, implicit in Step 2.

Number of A-solutions (no large prime) $=$ 538
Number of B-solutions (large prime) $=$ 1133

Step 3. Pair the large primes.

When the large prime q arises $l > 2$ times, we count $l - 1$ pairs.

Number of A-solutions $=$ 538
Pairs of B-solutions $=$ 399
Equations already known $=$ 81
Total $=$ 1018

Step 4. Obtain factorisations of $a + bx$, $a + bz$.

For those solutions to be used, whether A or B, obtain the complete factorisations of $a + bx$ by trial division (by working (mod p), we eliminate unsuccessful trials).

Factorising the numbers $[a, b, 0]$ is slightly more complicated. Again compute the norm $N(a, b, 0)$ and factor by trial division. When a prime p is found, divide out a prime of norm $\pm p$ from the number $[a, b, 0]$ (we may need to try up to three such primes). We should finish with a unit, $[d, e, f]$ say. From a table of powers of U and $1/U$, we can recognise it as $\pm U^i$.

Two questions arise here:
1. How large a table is needed? I took $i = -8 \ldots 8$, the largest I could compute in single length arithmetic (32 bits).
2. Perhaps this process could still cause an overflow? But it didn't. Numbers with small norm can have large coefficients (even units).

Step 5. Obtain linearly dependent sets.

Just as in QS (or CF). My QS program needed only trivial changes. 31 sets were obtained.

Step 6. Complete the factorisation.

Again just as in QS. Add up the powers of each member of FB_1 and FB_2 in all the solutions in the set (take the quotient of each pair of B-solutions). The totals should be even! Replacing each number $[a, b, c]$ by $\{a, b, c\}$, compute integers X and Y with

$$X^2 \equiv Y^2 \ (\text{mod } F_7).$$

Compute

$$\gcd(X - Y, F_7) = 59\,64958\,91274\,97217 \qquad (\text{1st set!}).$$

This agrees with [3], and with the well-known mnemonic for Fermat factors.

COMPUTER AND PROGRAM DETAILS

The 8-bit computer used is the Philips P2012, with 64k of store, and two disc drives (640k each). There are seven separate programs, one for each step described above. The programs are in Pascal, with a small amount of machine code (for multi-length arithmetic and logical operations). My programs have much in common with those written for QS (only six!), with which I have factored numbers up to 51 digits.

Program 1. Single length working. Rational primes are generated by a sieve. My method to find their factors in S is crude. Generate all numbers $[a, b, c]$ of S with small coefficients (I used $-15 \ldots 15$), saving those with norm ± 1 or $\pm p$. By experiment, find a fundamental unit U. Sort the others on p and remove associates to obtain FB_2.

Program 2. Similar to QS, e.g. [4]. I use a sieve array of 8 bit elements. Multi-length arithmetic not essential (only used if exact values of large primes are to be found at once).

Program 2A. Requires multilength working for $a + bx$ and double length working for the values of the norm. Inefficient on an 8-bit computer without division— unlike Program 2, which doesn't need them. Probably capable of improvement— but the author prefers to keep to Pascal if at all possible!

Program 3. Simple sorting program using Treesort3, CACM algorithm 245.

Program 4. Requires multilength arithmetic.

Program 5. Requires bit operations. Binary matrix is processed in a series of passes between two disc files (largest matrix used so far (for QS) was ~ 1700 rows).

Program 6. Requires multilength arithmetic and bit operations.

The running times quoted below are very rough since the programs have been run over a period of some weeks, and in some cases have gone through several versions.

Program	Time (hours)
1	1
2	5
2A	10
3	0.5
4	1.5
5	2
6	0.1
Total	20.1

Example of an A-solution

$$x = 2^{43}, \qquad a = 1693, \qquad b = 749.$$

$$a + bx = 6\,58827\,36736\,35485$$
$$= 5 \cdot 13 \cdot 19 \cdot 449 \cdot 1567 \cdot 2477 \cdot 3061$$

$$a^3 - 2b^3 = 40121\,80059$$
$$= \quad -1 \quad \cdot \quad -3 \quad \cdot \quad -43 \quad \cdot \quad 157 \quad \cdot \quad -397 \quad \cdot \quad 499$$

$$[a, b, 0] = [-1, -2, -1] \cdot [-1, 1, 0] \cdot [-3, 2, 0] \cdot [5, 0, 2] \cdot [-7, 3, 0] \cdot [5, 1, 4]$$

The first factor on the right is $-U^2$; the others are primes (written as in FB_2, with highest nonzero coefficient positive).

References

1. D. Coppersmith, A. M. Odlyzko, R. Schroeppel, *Discrete logarithms in $GF(p)$*, Algorithmica **1** (1986), 1–15.
2. I. N. Stewart, D. O. Tall, *Algebraic number theory*, second edition, Chapman and Hall, London, 1987.
3. M. Morrison, J. Brillhart, *A method of factoring and the factorization of F_7*, Math. Comp. **29** (1975), 183–205.
4. J. L. Gerver, *Factoring large numbers with a quadratic sieve*, Math. Comp. **41** (1983), 287–294.

TIDMARSH COTTAGE, MANOR FARM LANE, TIDMARSH, READING, BERKSHIRE, RG8 8EX, ENGLAND

THE NUMBER FIELD SIEVE

A. K. Lenstra, H. W. Lenstra, Jr., M. S. Manasse, J. M. Pollard

Abstract. The number field sieve is an algorithm to factor integers of the form $r^e - s$ for small positive r and $|s|$. The algorithm depends on arithmetic in an algebraic number field. We describe the algorithm, discuss several aspects of its implementation, and present some of the factorizations obtained. A heuristic run time analysis indicates that the number field sieve is asymptotically substantially faster than any other known factoring method, for the integers that it applies to. The number field sieve can be modified to handle arbitrary integers. This variant is slower, but asymptotically it is still expected to beat all older factoring methods.

1. Introduction

In this paper we present a novel algorithm to factor integers of the form $r^e - s$, where r and $|s|$ are small positive integers, $r > 1$, and e is large. The algorithm has become known as the *number field sieve*, because it depends on arithmetic in an algebraic number field combined with more traditional sieving techniques. It has proved to be quite practical, its most notable success being the factorization of the ninth Fermat number. We refer to our account [26] of the latter factorization for an introduction to the number field sieve.

Let N be an integer of the form $r^e - s$ as above. It should be thought of as an integer that we want to factor into prime factors. Examples of such N can be found in the Cunningham tables [3]. In many cases, one already knows some prime factors of N, so that it is the cofactor n that remains to be factored. Applying the number field sieve for this purpose is not recommended if n is much smaller than N, since the conjectured run time of the algorithm depends on the size of N rather than on the size of n.

To express the conjectured run time, we define

$$L_x[v, \lambda] = \exp(\lambda (\log x)^v (\log \log x)^{1-v})$$

for real numbers x, v and λ with $x > e$. In the discussion below we will, for simplicity, abbreviate the expression $L_x[v, \lambda + o(1)]$ to $L_x[v, \lambda]$; here the $o(1)$ is for $x \to \infty$. With this notation, we expect that for r and $|s|$ below a fixed upper bound the number field sieve takes time $L_N[\frac{1}{3}, c]$, where $c = (32/9)^{1/3} \doteq 1.5263$, irrespective of the size of the factors of N. We are not able to prove this run time rigorously, and even our heuristic argument has a weak spot (see 6.4).

1991 *Mathematics Subject Classification.* Primary 11Y05, 11Y40.

Key words and phrases. Factoring algorithm, algebraic number fields.

The second author was supported by NSF under Grant No. DMS–9002939 and by NSA/MSP under Grant No. MDA90–H–4043.

Buhler and Pomerance observed that the idea of the number field sieve can be applied to general integers as well, i.e., to integers n that do not necessarily have a small multiple of the form $r^e - s$ as above. The generalized version of the number field sieve is also conjectured to take time $L_n[\frac{1}{3}, c]$, but with a larger value for c than above. Due to additional contributions by Adleman and Coppersmith, the smallest value for c that can currently conjecturally be achieved is given by

$$c = \frac{(92 + 26\sqrt{13})^{1/3}}{3} \doteq 1.9019,$$

see Section 9, and [1; 7; 11]. This makes the number field sieve, conjecturally and asymptotically, into the fastest currently known integer factoring algorithm; only the elliptic curve method [29] is, for a special class of numbers, supposed to be faster.

The function $L_n[v, \lambda]$, which plays an important role in the analysis of modern factoring algorithms, interpolates between powers of n and powers of $\log n$. More specifically, we have

$$L_n[1, \lambda] = n^\lambda, \qquad L_n[0, \lambda] = (\log n)^\lambda,$$
$$\log \log L_n[v, \lambda] = v \cdot \log \log L[1, \lambda] + (1 - v) \cdot \log \log L[0, \lambda].$$

The most significant parameter is v. Traditional algorithms such as trial division have $v = 1$, in the sense that they run in time $L[1, \lambda]$ for some $\lambda > 0$. These algorithms are said to run in exponential time. A polynomial time algorithm would have $v = 0$. Thus, the many algorithms that run in time $L_n[\frac{1}{2}, \lambda]$ (see below) are in this sense halfway between the exponential time and polynomial time ones. The number field sieve, which has $v = \frac{1}{3}$, represents an additional step in the direction of polynomial time algorithms. The notation $L_n[v, \lambda]$ was introduced in [25], following the notation $L(n)^\lambda$ that Pomerance introduced for $L_n[\frac{1}{2}, \lambda]$ in 1983 (see [34]).

The first person to realize that the function $L_n[\frac{1}{2}, \lambda]$ can be used to express conjectural run times of factoring algorithms was Schroeppel, in 1975 (see [21, Section 4.5.4]). In 1978 Dixon obtained a rigorous result of this nature (see [16]). The study of the precise value of λ was initiated by Pomerance (see [34]). It is now conjectured that many factoring algorithms, including the continued fraction method, the quadratic sieve, and the elliptic curve method, run in expected time at most $L_n[\frac{1}{2}, 1]$, and for the class group relations method this has been proved (see [31]). Accounts of these developments can be found in [25] and [36]. The *cubic sieve* algorithm (see [14, Section 7; 25, Section 4.E]) is conjectured to be faster and to run in time $L_n[\frac{1}{2}, c']$, with $\sqrt{2/3} \le c' < 1$; however, it applies only to numbers of a special form, including Cunningham numbers, and it has never proved to be of more than theoretical interest.

In 1981 the function $L_n[\frac{1}{3}, \lambda]$ made its appearance in the analysis of factoring algorithms, when Schnorr [38] showed, under plausible assumptions, that an integer n can be factored in time $L_n[\frac{1}{3}, 2]$ provided that suitable lists of *smooth numbers* are available. Here we call a number *smooth*—the term is due to Rivest—if its prime factors are small; more precisely, it is *B-smooth* if its prime factors are at most B.

The importance of the function $L_n[v, \lambda]$ for analyzing factoring algorithms is due to its connection with smooth numbers. Many factoring algorithms proceed by generating a sequence of integers, in a more or less random fashion, of which only the smooth ones are useful. For all algorithms before the number field sieve, the integers that are inspected for smoothness have order of magnitude n^w, for some constant w that depends on the algorithm. For instance, all algorithms with expected run time $L_n[\frac{1}{2}, 1]$ have $w = \frac{1}{2}$. The cubic sieve algorithm was the first to break through the $w = \frac{1}{2}$ barrier with $w = \frac{1}{3}$ for n of a special form, and thus achieved run time $L_n[\frac{1}{2}, c']$ with $c' < 1$ for such n. A more dramatic improvement is realized by the number field sieve: the integers that it inspects for smoothness are only $n^{o(1)}$, if n is not too much smaller than N. This makes it the first factoring algorithm with conjectured run time essentially faster than $L_n[\frac{1}{2}, \lambda]$, for any positive constant λ. We refer to [7, Section 10] for a discussion of the relation between the run time of a factoring algorithm and the size of the numbers that it inspects for smoothness, and for an explanation of the role played by the function $L_n[v, \lambda]$.

From an algorithmic point of view, the problem of factoring integers is closely related to the discrete logarithm problem, see [25; 36]. The conjectural run time of many discrete logarithm algorithms for a finite field of n elements is of the form $L_n[v, \lambda]$ with $v = \frac{1}{2}$. Coppersmith was the first to achieve $v = \frac{1}{3}$, for the case that n is a power of 2 (see [10; 25, Section 3.17; 19]). His *bimodal polynomials* method shares a few formal features with Schnorr's work [38]: the appearance of $v = \frac{1}{3}$, and the requirement that two expressions are simultaneously smooth. It is interesting to observe that the number field sieve has these features as well. In addition, both the number field sieve and the bimodal polynomials method start by looking for a good auxiliary polynomial.

The algorithm in the present paper was inspired by the discrete logarithm algorithm for prime n using Gaussian integers that was presented by Coppersmith, Odlyzko, and Schroeppel (see [14, Section 7; 23]). This algorithm, which has $v = \frac{1}{2}$, was in turn inspired by work of ElGamal [17]. The main change that we made, which is crucial for obtaining $v = \frac{1}{3}$, is that we use rings of algebraic integers in higher degree number fields, and that we optimize the choice of the degree as a function of the number to be factored (see 6.3). Gordon [18] showed that the same technique can be used for the discrete logarithm problem for prime n. The conjectural run time estimate of his algorithm is $L_n[\frac{1}{3}, 9^{1/3}]$, where $9^{1/3} \doteq 2.0801$. Schirokauer [39] improved this to $L_n[\frac{1}{3}, (64/9)^{1/3}]$, where $(64/9)^{1/3} \doteq 1.9230$.

In Section 2 of the present paper we describe the number field sieve, as it applies to integers of the form $r^e - s$ for small positive r and $|s|$. Details that are left out from this description are explained in Sections 3, 4, and 5. A conjectural analysis of the run time of the number field sieve is given in Section 6. In Section 7 we discuss a few modifications to the algorithm. Examples of factorizations that have been obtained by means of the number field sieve are presented in Section 8.

Section 9 is devoted to possible generalizations of the number field sieve to arbitrary integers.

In our description of the number field sieve we will make a few simplifying and not always realistic assumptions about the number fields that we are using, cf. 2.5 and 3.1. We refer to 3.4–3.8, Section 9, and [2; 7] for variations of the number field sieve that make no simplifying assumptions about the fields involved.

We shall denote by $\mathbf{Z}$ the ring of integers, and by $\mathbf{Q}$, $\mathbf{R}$, and $\mathbf{C}$ the fields of rational, real, and complex numbers, respectively.

2. The algorithm

Let n be an odd integer, $n > 1$, and assume that n is not a prime number or a power of a prime number. We assume that n itself or a small multiple of n is of the form $r^e - s$, for a small integer $r > 1$ and a non-zero integer s of small absolute value, and with e an integer that is possibly much larger. It is assumed that r, e, and s are given along with n. Numbers of this form often appear on the 'wanted' lists from [3]. The fact that n is not a prime number can usually easily be proved by means of a probabilistic compositeness test, see [25, Section 5.1]. If we use the variation of the probabilistic compositeness test described in [26, Section 2.5] we can also easily check that n is not a prime power. We describe a factoring algorithm, the *number field sieve*, that makes use of the special form of the multiple $r^e - s$ of n, to factor n. For background on the elementary algebraic number theory used by the algorithm we refer to [40] and to [26, Section 4].

2.1. *Outline of the algorithm.* For a random integer x satisfying

$$(2.2) \qquad x^2 \equiv 1 \bmod n$$

there is a probability of at least $\frac{1}{2}$ that $\gcd(n, x - 1)$ is a non-trivial factor of n. To factor n it therefore suffices to construct several solutions x to (2.2) in an apparently random manner. Many factoring algorithms, including the number field sieve, achieve this by means of the following three-step approach.

Step 1. *Selecting the factor base.* Select a collection of non-zero elements $a_i \in \mathbf{Z}/n\mathbf{Z}$, with i ranging over some finite index set I. The collection $(a_i)_{i \in I}$ is called the *factor base*, for a reason that will be clear from the sequel. The elements of the factor base should not be confused with a possible list of candidate factors of n; indeed, we may assume that all a_i are units in $\mathbf{Z}/n\mathbf{Z}$, because if they are not, then n can be factored immediately.

Step 2. *Collecting relations.* Collect *relations* between the a_i, i.e., vectors $v = (v_i)_{i \in I} \in \mathbf{Z}^I$ for which

$$(2.3) \qquad \prod_{i \in I} a_i^{v_i} = 1.$$

Stop as soon as the collection V of relations that have been found contains slightly more than $\#I$ elements.

Step 3. *Finding dependencies.* Find dependencies modulo 2 among the elements of V, i.e., subsets W of V such that $\sum_{v \in W} v = 2 \cdot (w_i)_{i \in I}$ with $w_i \in \mathbf{Z}$.

Notice that non-trivial dependencies exist because $\#V > \#I$. For each dependency W we can calculate an integer x with $\prod_{i \in I} a_i^{w_i} = (x \bmod n)$, and this integer satisfies (2.2). Under conditions on the a_i and V that can usually not be proved but that are normally satisfied, consideration of a few linearly independent W's leads to the complete factorization of n into powers of distinct prime numbers; see [26, Section 2.6] for a further discussion of this point.

The remainder of this section is devoted to a description of how Steps 1 and 2 are carried out in the number field sieve. For Step 3 we refer to the literature on large sparse matrix elimination, cf. [22; 37; 42; 12; 13], and to [26].

2.4. *The idea of the number field sieve.* The number field sieve is based on the observation that it is possible to construct a number field $K = \mathbf{Q}(\alpha)$ and a ring homomorphism φ from the subring $\mathbf{Z}[\alpha]$ of K to $\mathbf{Z}/n\mathbf{Z}$ such that $\varphi(\alpha) = (m \bmod n)$, where both α and $|m|$ are small compared to n; here the smallness of α is measured by means of the sum of the absolute values of the coefficients of its irreducible polynomial, which are supposed to be integers. The idea is then to look for pairs of small coprime integers a and b such that both the algebraic integer $a + b\alpha$ and the integer $a + bm$ are smooth, in a sense to be specified below. Because $\varphi(a + b\alpha) = (a + bm \bmod n)$, each pair provides an equality of two products in $\mathbf{Z}/n\mathbf{Z}$. The factors occurring in these products form the factor base, and each congruence leads to a relation as in (2.3).

2.5. *Construction of the number field.* To define the number field K and the homomorphism φ as in 2.4 we proceed as follows. Given the multiple $r^e - s$ of n, we first select a small positive integer d that will serve as the extension degree. More about the choice of d can be found in Sections 6, 7, and 8. Given d, let k be the least positive integer for which $k \cdot d \geq e$, put $t = s \cdot r^{k \cdot d - e}$, and let f be the polynomial $X^d - t$. The number $m = r^k$ satisfies $f(m) \equiv 0 \bmod n$, since n divides $r^e - s$. Our number field K is now given by $K = \mathbf{Q}(\alpha)$, where $f(\alpha) = 0$.

We will assume that the polynomial f is irreducible. This condition is likely to be satisfied, since in realistic cases a non-trivial factor of f gives rise to a non-trivial factor of n. If it is not satisfied we can replace f by a suitable factor. The irreducibility of f is easily checked: f is reducible if and only if either there is a prime number p dividing d such that t is a pth power, or 4 divides d and $-4t$ is a fourth power (see [24, Chapter VI, Theorem 9.1]). For example, if r is not a power of a smaller integer, and $s = 1$, then f is irreducible if and only if $\gcd(d, e) = 1$.

The irreducibility of f implies that the degree of the number field K equals d, and that each element of K has a unique expression of the form $\sum_{i=0}^{d-1} q_i \alpha^i$ with $q_i \in \mathbf{Q}$. The subring $\mathbf{Z}[\alpha]$ of K consists of the expressions $\sum_{i=0}^{d-1} s_i \alpha^i$ with coefficients $s_i \in \mathbf{Z}$. The ring homomorphism $\varphi \colon \mathbf{Z}[\alpha] \to \mathbf{Z}/n\mathbf{Z}$ is now defined by $\varphi(\alpha) = (m \bmod n)$. Generally, we have $\varphi(\sum_{i=0}^{d-1} s_i \alpha^i) = (\sum_{i=0}^{d-1} s_i m^i \bmod n)$ if $s_i \in \mathbf{Z}$.

To simplify the exposition of the algorithm, we will assume that *the ring $\mathbf{Z}[\alpha]$ is a unique factorization domain.* As we shall see in 3.4, this is a strong assumption, which is not always satisfied. We refer to Section 3 for a discussion of the modifications that are necessary if $\mathbf{Z}[\alpha]$ is not a unique factorization domain.

To give an example, for $3^{239}-1$ (one of the numbers we factored, cf. Section 8), we used $d = 5$ as extension degree, $m = 3^{48}$, and α a zero of the polynomial $f = X^5 - 3$; in this case $\mathbf{Z}[\alpha]$ is indeed a unique factorization domain. For another number we factored, $2^{512} + 1$, we used $d = 5$, $m = 2^{103}$, and α a zero of $X^5 + 8$. Because $\alpha^2/2 \notin \mathbf{Z}[\alpha]$ and because $\alpha^2/2$ is a zero of $X^5 - 2 \in \mathbf{Z}[X]$, we find that $\mathbf{Z}[\alpha]$ is not the ring of integers of K. We simply got around that problem by using $\mathbf{Z}[\alpha^2/2]$ instead of $\mathbf{Z}[\alpha]$ in the algorithms described below.

It is in the construction of K and φ, as described above, that we exploit the special form of the multiple $r^e - s$ of n. The main difficulty with general n is that one is led to consider much "larger" number fields, which are much harder to control. This difficulty is discussed in Section 9.

2.6. *Smoothness in the number field.* An algebraic integer is called *B-smooth* if every prime number dividing its norm is at most B. We shall mainly be interested in smoothness of algebraic integers of the form $a + b\alpha$, where a, b are coprime integers. The norm $\mathbf{N}(a + b\alpha)$ of $a + b\alpha$ is equal to $a^d - t(-b)^d$, so $a + b\alpha$ is B-smooth if and only if $|a^d - t(-b)^d|$ is a product of prime numbers $\le B$.

The *norm* $\mathfrak{N}\mathfrak{a}$ of a non-zero ideal $\mathfrak{a}$ of $\mathbf{Z}[\alpha]$ is defined by $\mathfrak{N}\mathfrak{a} = \#(\mathbf{Z}[\alpha]/\mathfrak{a})$, which is a positive integer. A *first degree prime ideal* of $\mathbf{Z}[\alpha]$ is a non-zero ideal $\mathfrak{p}$ of prime norm p. For such an ideal we have $\mathbf{Z}[\alpha]/\mathfrak{p} \cong \mathbf{Z}/p\mathbf{Z}$, which is a field, so that $\mathfrak{p}$ is indeed a prime ideal. The set of first degree prime ideals $\mathfrak{p}$ is in bijective correspondence with the set of pairs $(p, c \bmod p)$, where p is a prime number and $c \in \mathbf{Z}$ satisfies $f(c) \equiv 0 \bmod p$; if $\mathfrak{p}$ corresponds to $(p, c \bmod p)$, then $\mathfrak{N}\mathfrak{p} = p$, the map $\mathbf{Z}[\alpha] \to \mathbf{Z}[\alpha]/\mathfrak{p} \cong \mathbf{Z}/p\mathbf{Z}$ maps α to $(c \bmod p)$, and $\mathfrak{p}$ is generated, as an ideal, by p and $c - \alpha$. The map $\mathbf{Z}[\alpha] \to \mathbf{Z}[\alpha]/\mathfrak{p}$ can be used to test whether a given element of $\mathbf{Z}[\alpha]$ is contained in $\mathfrak{p}$: namely, one has $\sum_i s_i\alpha^i \in \mathfrak{p}$ if and only if $\sum_i s_i c^i \equiv 0 \bmod p$, with p, c as above.

Let a, b be coprime integers. Every prime ideal of $\mathbf{Z}[\alpha]$ that contains $a + b\alpha$ is a first degree prime ideal (see [26, Section 5, Lemma; 7, Corollary 5.5]), and as we just saw $a + b\alpha$ is contained in the prime ideal corresponding to $(p, c \bmod p)$ if and only if $a + bc \equiv 0 \bmod p$. This implies that the prime ideal factorization of $a + b\alpha$ corresponds to the prime factorization of its norm $a^d - t(-b)^d$, as follows. If $a^d - t(-b)^d$ contains the prime factor p exactly k times, with $k > 0$, then $a \equiv -bc \bmod p$ for a unique $c \bmod p$ for which $f(c) \equiv 0 \bmod p$, and the first degree prime ideal corresponding to $(p, c \bmod p)$ divides $a + b\alpha$ exactly to the kth power. So, one ideal of norm p accounts for the full exponent of p in $a^d - t(-b)^d$.

We denote by $\pi_\mathfrak{p}$ an element of $\mathbf{Z}[\alpha]$ that generates $\mathfrak{p}$. Such an element exists because $\mathbf{Z}[\alpha]$ is assumed to be a principal ideal domain; it is unique only up to multiplication by units. To pass from the prime ideal factorization of $a + b\alpha$ to its prime factorization it suffices to replace each prime ideal factor $\mathfrak{p}$ by $\pi_\mathfrak{p}$ and to multiply the result by a suitable unit.

2.7. *Step 1 of the number field sieve.* The discussion above leads to the following selection of the factor base. First select two smoothness bounds, B_1 and B_2. In practice these bounds are best determined empirically. See Section 8 for some examples, and 6.2 and 6.3 for choices that are satisfactory from a theoretical point of view. We will use B_1 as smoothness bound for the integers $a + bm$, and B_2 as smoothness bound for the algebraic integers $a + b\alpha$. Now let $I = P \cup U \cup G$, where P is the set of all prime numbers $\leq B_1$, the set U is a set of generators for the group of units of $\mathbf{Z}[\alpha]$, and G consists of the elements $\pi_\mathfrak{p} \in \mathbf{Z}[\alpha]$, where $\mathfrak{p}$ ranges over the set of first degree prime ideals of $\mathbf{Z}[\alpha]$ of norm $\leq B_2$. The factor base is then formed by the elements $a_i = \varphi(i) \in \mathbf{Z}/n\mathbf{Z}$, for $i \in I$. We assume that $\gcd(a_i, n) = 1$ for $i \in I$; if that is not the case n can easily be factored, and the algorithm terminates.

To complete the description of Step 1 it remains to explain how to construct the sets U and G. This will be done in Section 3.

2.8. *Step 2 of the number field sieve.* We discuss how Step 2 is performed. First, select two additional bounds B_3 and B_4. These bounds are again best determined empirically. See 4.6 for various considerations concerning their choice, and Section 8 for examples. To find relations among the a_i, one searches for pairs of integers (a, b), with $b > 0$, satisfying the following conditions:

(i) $\gcd(a, b) = 1$;

(ii) $|a + bm|$ is B_1-smooth, except for at most one additional prime factor p_1, which should satisfy $B_1 < p_1 < B_3$;

(iii) $a + b\alpha$ is B_2-smooth, except for at most one additional prime ideal factor $\mathfrak{p}_2$, of which the norm p_2 should satisfy $B_2 < p_2 < B_4$.

We will assume that $a + bm > 0$; in the unlikely event that $a + bm < 0$, replace (a, b) by $(-a, -b)$.

The prime number p_1 in (ii) is called the *large prime*, and the additional prime ideal $\mathfrak{p}_2$ in (iii) the *large prime ideal*. Note that $\mathfrak{p}_2$ corresponds to the pair $(p_2, c \bmod p_2)$, where c is such that $a \equiv -bc \bmod p_2$; this enables us to distinguish between prime ideals of the same norm. If the large prime does not occur, then we write $p_1 = 1$. Likewise, if the large prime ideal does not occur in (iii), we write symbolically $\mathfrak{p}_2 = 1$ and $p_2 = 1$. Pairs (a, b) for which $p_1 = p_2 = 1$ will be called *full* relations, and the other pairs *partial* relations.

The search for pairs (a, b) satisfying (i), (ii), and (iii) above can be carried out by means of the sieving technique described in Section 4. We show how the pairs give rise to relations between the a_i. First, suppose that (a, b) is a full relation. By (ii), there is an identity of the form

$$a + bm = \prod_{p \in P} p^{e(p)},$$

with $e(p) \in \mathbf{Z}_{\geq 0}$. From (i), (iii), and 2.6 it follows that $a + b\alpha$ can be written as a product of elements $\pi_\mathfrak{p} \in G$ to certain powers, and a unit from $\mathbf{Z}[\alpha]$. One can determine the contribution from G by considering the factorization of $N(a + b\alpha) = a^d - t(-b)^d$, as explained in 2.6. Since U generates the group of units of $\mathbf{Z}[\alpha]$, the unit contribution to the prime factorization of $a + b\alpha$ can be written as a product of elements from U. In Section 5 it is explained how this is achieved. As a result, we get an identity of the form

$$a + b\alpha = \prod_{u \in U} u^{e(u)} \cdot \prod_{g \in G} g^{e(g)},$$

with $e(u) \in \mathbf{Z}$ and $e(g) \in \mathbf{Z}_{\geq 0}$. Because $a + bm$ and $a + b\alpha$ have the same image under φ, these two factorizations lead to the identity

(2.9)
$$\prod_{p \in P} \varphi(p)^{e(p)} = \prod_{u \in U} \varphi(u)^{e(u)} \cdot \prod_{g \in G} \varphi(g)^{e(g)}$$

in $\mathbf{Z}/n\mathbf{Z}$, from which one obtains a relation $v = (v_i)_{i \in I} \in \mathbf{Z}^I$ between the a_i by putting $v_i = e(i)$ for $i \in P$, and $v_i = -e(i)$ for $i \notin P$. Note that $a_i^{v_i}$ exists for $i \notin P$ because $\gcd(a_i, n) = 1$ for $i \in I$.

In this way each full relation (a, b) leads to a relation between the a_i as in (2.3).

2.10. *Making use of partial relations.* As we will see in Section 4, partial relations can be found at little extra cost during the search for full relations. Furthermore, they occur much more frequently than full relations, so that relatively many of them are found; so many, in fact, that there are quite a few with the same large prime or the same large prime ideal. If that occurs it may be possible to convert a collection of partial relations into a relation among the a_i as in (2.3), as follows.

A set C of partial relations is called a *cycle* if for each $(a, b) \in C$ there is a sign $s(a, b) \in \{-1, +1\}$ such that

$$\prod_{(a,b) \in C} (a + bm)^{s(a,b)} = \prod_{p \in P} p^{e(p)},$$

with $e(p) \in \mathbf{Z}$, and

$$\prod_{(a,b) \in C} (a + b\alpha)^{s(a,b)} = \prod_{u \in U} u^{e(u)} \cdot \prod_{g \in G} g^{e(g)},$$

with $e(u), e(g) \in \mathbf{Z}$. Informally, this means that if a prime $p_1 > B_1$ occurs in the factorization of $a + bm$ for some pair $(a, b) \in C$, then p_1 also occurs in the factorization of $\bar{a} + \bar{b}m$ for some pair $(\bar{a}, \bar{b}) \in C$ with $s(\bar{a}, \bar{b}) = -s(a, b)$. Similarly, each occurrence of a large prime ideal in $a + b\alpha$ is canceled by the occurrence of the same large prime ideal in another pair with the opposite sign.

For each cycle one can compute the exponents $e(p)$ by adding or subtracting the exponents occurring in the factorizations of the $a + bm$ for the pairs (a, b) in the cycle, according to their signs. Similarly, one computes the $e(g)$ by adding or subtracting the exponents in the prime ideal factorizations of the $a + b\alpha$; these exponents are found as explained in 2.6. Once the $e(g)$ have been computed, the $e(u)$ are found with the method given in Section 5, cf. Remark 5.3.

Just as above, we now obtain the following relation between the a_i:

$$\prod_{p \in P} \varphi(p)^{e(p)} = \prod_{u \in U} \varphi(u)^{e(u)} \cdot \prod_{g \in G} \varphi(g)^{e(g)}.$$

This is the same as (2.9), except that now the integers $e(i)$ are allowed to be negative. Since $\gcd(a_i, n) = 1$ for $i \in I$, negative powers of the a_i are well defined.

In this way each cycle among partial relations leads to a relation between the a_i.

2.11. *Remark.* Partial relations for which $p_1 \neq 1$ but $p_2 = 1$ are referred to as *pf*'s (for 'partial-full'), because they would lead to an equation as in (2.9) with a partial factorization, i.e., a factor $\varphi(p_1)$, on the left hand side, and a full factorization on the right. Similarly, partial relations for which $p_1 = 1$ and $p_2 \neq 1$ are called *fp*'s, partial relations for which both $p_1 \neq 1$ and $p_2 \neq 1$ are called *pp*'s, and full relations are *ff*'s. We refer to 7.3 for a more general notion of partial relations.

The negative $s(a, b)$ in the cycles have the effect that the large primes and the large prime ideals in the resulting combinations are canceled. For cycles consisting of only two *pf*'s (with the same p_1) there is no need to make use of the signs $s(a, b)$, because a relation among the a_i of the form

$$x^2 \cdot \prod_{i \in I} a_i^{v_i} = 1$$

for some unit $x \in \mathbf{Z}/n\mathbf{Z}$ (of the form $\varphi(p_1)$), is just as useful as a relation like (2.3). However, doing the same for cycles involving *fp*'s or *pp*'s would introduce factors $\varphi(\pi_{\mathfrak{p}})$ for prime ideals $\mathfrak{p}$ of norm $> B_2$ into x, and would thus require finding generators of the large prime ideals involved in the cycles. This is avoided by means of the signs described above.

In 9.6 we shall encounter a variant of the number field sieve in which the use of generators is avoided, and in which the signs can be discarded. At the other extreme, 9.1 describes a variant in which *all* prime ideals are canceled, not just the large ones.

2.12. *Finding the cycles.* Write P_1 for the set of all large primes occurring in the partial relations, and P_2 for the set of all large prime ideals that occur. We view the set of partial relations as the set of edges of a graph with vertex set $\{1\} \cup P_1 \cup P_2$; namely, each partial relation with large prime p_1 and large prime ideal $\mathfrak{p}_2$ represents an edge between p_1 and $\mathfrak{p}_2$. The edges incident with 1 correspond to the *pf*'s and *fp*'s; except for these edges the graph is bipartite. Each cycle in the graph gives rise to a cycle among the partial relations. For cycles of even length one can assign the signs $+1$, -1 alternately to the partial relations corresponding to the edges. Cycles of odd length contain the vertex 1; again one can assign the signs alternately, but now starting with an edge incident with 1.

It is not necessary to find *all* cycles in the graph. For example, if the symmetric difference of two cycles C_1 and C_2 is a cycle C_3, then the relation between the

a_i obtained from C_3 is a linear combination of the relations obtained from C_1 and C_2. Therefore, if the cycles C_1, C_2 are already used, there is no point in using C_3 as well. In other words, it will suffice to find a maximal set of cycles that is "independent" in a suitable sense. In [28] it is explained how this can be done, and how a convenient representation for the graph can be constructed. Another way of dealing with the partial relations is to postpone their combination into cycles until Step 3, as discussed in 7.2.

2.13. *Free relations.* In addition to the relations that are based on full relations and cycles among partial relations, there are the *free* relations, which are much easier to come by. They are already valid in the ring $\mathbf{Z}[\alpha]$, before φ is applied. There is one such relation for each prime number $p \leq \min(B_1, B_2)$ for which the polynomial $f = X^d - t$ factors completely into linear factors modulo p. Namely, let p be such a prime, and write $X^d - t \equiv \prod_c (X - c)^{e_c} \bmod p$, where c ranges over a set of integers that are pairwise distinct modulo p, and where the multiplicities e_c are positive integers (they are equal to 1 if p does not divide dt). Each c is a zero of $f \pmod{p}$, and therefore gives rise to a first degree prime ideal $\mathfrak{p}$ of norm p, as explained in 2.6; for this $\mathfrak{p}$ we write $e(\mathfrak{p}) = e_c$. With this notation, the ideal generated by p is equal to the product of the ideals $\mathfrak{p}^{e(\mathfrak{p})}$, so p can be written as the product of the elements $\pi_{\mathfrak{p}}^{e(\mathfrak{p})}$ multiplied by a unit:

$$p = \prod_{u \in U} u^{e(u)} \cdot \prod_{\mathfrak{p}, \, \mathfrak{N}\mathfrak{p}=p} \pi_{\mathfrak{p}}^{e(\mathfrak{p})} \qquad \text{with } e(u) \in \mathbf{Z}.$$

This gives the identity

$$\varphi(p) = \prod_{u \in U} \varphi(u)^{e(u)} \cdot \prod_{\mathfrak{p}, \, \mathfrak{N}\mathfrak{p}=p} \varphi(\pi_{\mathfrak{p}})^{e(\mathfrak{p})},$$

which has the form (2.9). The integers $e(u) \in \mathbf{Z}$ can again be found by means of the method explained in Section 5. The density of the set of primes that split completely in this way is the inverse of the degree of the splitting field of f, which divides $d \cdot \phi(d)$ and is a multiple of $\mathrm{lcm}(d, \phi(d))$, with ϕ the Euler ϕ-function. For example, if $d = 5$ then one out of every twenty primes splits completely in this way; if $B_1 \approx B_2$ this means that one may expect approximately one fortieth of all relations to come for free.

This completes the description of Step 2, and thereby the description of the number field sieve.

3. FINDING GENERATORS

In this section we discuss the computation of the sets U and G introduced in 2.7.

An element of $\mathbf{Z}[\alpha]$ is a unit if and only if it has norm ± 1. The structure of the group of units can be described as follows. Let the polynomial f that was selected in 2.5 have r_1 real roots and $2r_2$ non-real complex roots, so that $r_2 = (d - r_1)/2$. Since f is of the form $X^d - t$, we have $r_1 = 1$ if d is odd; and if d is even, then $r_1 = 0$ if $t < 0$ and $r_1 = 2$ if $t > 0$. Let $l = r_1 + r_2 - 1$. With this notation, the group of units of $\mathbf{Z}[\alpha]$ is generated by a suitable root of unity u_0 and l multiplicatively independent units $u_1, u_2, \ldots, u_l$ of infinite order; we may take $u_0 = -1$ if $r_1 > 0$. We shall let U consist of such elements $u_0, \ldots, u_l$.

Before we compute G we make a list of all first degree prime ideals of norm $\leq B_2$. As we saw in 2.6, this amounts to making a list of all pairs $(p, c \bmod p)$, where p is a prime number $\leq B_2$, and $c \in \mathbf{Z}$ satisfies $f(c) \equiv 0 \bmod p$. To find these pairs efficiently, one can use a probabilistic root finder for polynomials over finite fields, cf. [21, Section 4.6.2]; the number of pairs thus found, i.e., $\#G$, can be expected to be close to $\pi(B_2)$, the number of primes up to B_2. An element of $\mathbf{Z}[\alpha]$ generates $\mathfrak{p}$ if and only if it belongs to $\mathfrak{p}$ and has norm $\pm p$; in other words, the conditions to be met by $\pi_{\mathfrak{p}} = \sum_{i=0}^{d-1} s_i \alpha^i$ are that $\sum_{i=0}^{d-1} s_i c^i \equiv 0 \bmod p$ and $\mathbf{N}(\pi_{\mathfrak{p}}) = \pm p$. To determine G it suffices to find one such element for each pair $(p, c \bmod p)$.

In practice the search for elements of U and G is best carried out simultaneously. This can be done as follows. Fix a multiplier bound M and a search bound C, depending on K and B_2. We refer to 3.6 for a discussion of feasible choices of M and C. For the moment, one may think of M as a fairly small integer—in all cases that we did M could be taken less than 10—and of C as roughly proportional to $B_2^{2/d}$. The actual asymptotics are a little different, though.

For all first degree prime ideals $\mathfrak{p}$ for which we want to find a generator, put $m(\mathfrak{p})$ equal to $M + 1$. This number $m(\mathfrak{p})$ keeps track of the status of $\mathfrak{p}$ during the search process: if $m(\mathfrak{p}) > M$ no generator has been found yet, otherwise an element $\bar{\pi}_{\mathfrak{p}}$ of $\mathfrak{p}$ has been found with $\mathbf{N}(\bar{\pi}_{\mathfrak{p}}) = \pm m(\mathfrak{p})p$, where $p = \mathfrak{N}\mathfrak{p}$; then the ideal generated by $\bar{\pi}_{\mathfrak{p}}$ is $\mathfrak{p}$ times an ideal of norm $m(\mathfrak{p})$.

3.1. *Search algorithm.* For all $\gamma = \sum_{i=0}^{d-1} s_i \alpha^i \in \mathbf{Z}[\alpha]$ for which $\sum_{i=0}^{d-1} s_i^2 |\alpha|^{2i} \leq C$, compute the norm $\mathbf{N}(\gamma)$, cf. Remark 3.3; here $|\alpha|$ denotes the real number $|t|^{1/d}$. If $\mathbf{N}(\gamma)$ is of the form kp for some prime p from the list of pairs $(p, c \bmod p)$ and some non-zero integer k with $|k| \leq M$, do the following. Identify the first degree prime ideal $\mathfrak{p}$ that corresponds to this p and γ, in other words, the pair $(p, c \bmod p)$ for which $\sum_{i=0}^{d-1} s_i c^i \equiv 0 \bmod p$, and update the data concerning $\mathfrak{p}$: if $m(\mathfrak{p}) > |k|$ then replace $m(\mathfrak{p})$ by $|k|$ and put $\bar{\pi}_{\mathfrak{p}}$ equal to γ.

After all these γ have been processed, the $m(\mathfrak{p})$ are all $\leq M$ if the multiplier bound M and the search bound C have been chosen properly. For the $\mathfrak{p}$ with $m(\mathfrak{p}) = 1$, put $\pi_{\mathfrak{p}}$ equal to $\bar{\pi}_{\mathfrak{p}}$. For the other $\mathfrak{p}$, compute $\pi_{\mathfrak{p}}$ by dividing $\bar{\pi}_{\mathfrak{p}}$ by a generator of the appropriate ideal of norm $m(\mathfrak{p})$. This requires the computation of generators of the ideals of norm at most M, as well as the inverses of these generators. There are only a few such ideals, and generators for them are often easy to find; in general, one may hope to encounter them during the search just described. If this doesn't work, one may have to appeal to one of the methods indicated in 3.8 below.

During the same search one keeps track of the units that are encountered. These are not only the elements γ with $\mathbf{N}(\gamma) = \pm 1$ that are found, but also quotients of two elements that have the same norm (up to sign) and that generate the same ideal; in the latter case a division is needed. Multiplicative dependencies can be cast out with the help of the function ν from Section 5. The set U of units that we are left with will often be the set of l multiplicatively independent elements that we are looking for. If later in the algorithm it is discovered that the resulting set U does not generate the group of units of $\mathbf{Z}[\alpha]$, then this discovery leads to a new unit, which can be used to alter U; see Remark 5.4.

3.2. *Remark.* If $r_1 > 0$ it is useful to require that all elements of G and U except $u_0 = -1$ are positive under some particular embedding of $\mathbf{Q}(\alpha)$ into $\mathbf{R}$. For this purpose one fixes one particular real embedding, and one replaces x by $-x$ for each $x \in G \cup U$, $x \neq -1$ that is negative under this embedding.

3.3. *Remark.* For $\gamma = \sum_{i=0}^{d-1} s_i \alpha^i$ the norm $\mathbf{N}(\gamma)$ is a homogeneous dth degree polynomial in the s_i with coefficients that are integers depending on the polynomial f. In our implementation of Algorithm 3.1 the norm-polynomial was 'hard-wired', i.e., each new f required changes in the program and thus recompilation. Furthermore, the search was organized in such a way that the norm of each γ was obtained from the norm of the previous γ by just a few arithmetic operations. This greatly enhanced the speed of our searching program.

In the rest of this section we discuss a few technical difficulties related to the search for U and G. Some of these difficulties were actually encountered during the factorizations reported in Section 8. Some others we did not encounter, but we can vividly imagine that this will happen to others who try the algorithm. Finally, there are difficulties of a primarily theoretical nature, which come up when one attempts to analyse the run time of the algorithm. We resolve these difficulties by using the tools that have been developed in algorithmic algebraic number theory. For general background, see [43; 8; 30; 33].

3.4. *Lack of unique factorization.* In the description of the algorithm we made the assumption that $\mathbf{Z}[\alpha]$ is a unique factorization domain. This is a strong assumption, which indeed fails to hold in three of the examples given in Section 8. The assumption implies that $\mathbf{Z}[\alpha]$ is equal to the ring of integers of K, which in turn implies that $e \equiv 0$ or $-1 \bmod d$. We now discuss how to proceed if $\mathbf{Z}[\alpha]$ is not assumed to have unique factorization. We note that in general it is not easy to check whether $\mathbf{Z}[\alpha]$ has unique factorization, but if desired this can be done along the way.

3.5. *The ring of integers.* One starts by replacing $\mathbf{Z}[\alpha]$ by the ring A of algebraic integers in K. Methods for determining A can be found in the references just given; see also [5]. The discriminant of f, which equals $\pm d^d t^{d-1}$, can for bounded r and $|s|$ easily be factored into primes, and once this prime factorization is available the determination of A proceeds in time $(d + \log |t|)^{O(1)}$. A few examples of rings of integers A are given in 2.5 and in Section 8. If $\mathbf{Z}[\alpha] \neq A$ then $\mathbf{Z}[\alpha]$ is not a unique factorization domain.

We shall denote the absolute value of the discriminant of A by Δ. This number is given by $\Delta = d^d |t|^{d-1} / [A : \mathbf{Z}[\alpha]]^2$, and it is usually determined simultaneously with A. One can show that Δ divides $d^d (r|s|)^{d-1}$ and that it is at least $d^d / (1 + \log d)^{cd}$ for some absolute positive constant c.

The replacement of $\mathbf{Z}[\alpha]$ by A necessitates a few modifications to the algorithm. The first is that the ring homomorphism $\varphi \colon \mathbf{Z}[\alpha] \to \mathbf{Z}/n\mathbf{Z}$ needs to be extended to A. This can be done if the natural condition $\gcd(drs, n) = 1$ is satisfied. Namely, any element $\gamma \in A$ can be written as $\gamma = \beta/m$, where $m \in \mathbf{Z}$ is built up from prime numbers dividing drs; then $\varphi(m)$ has an inverse in $\mathbf{Z}/n\mathbf{Z}$, and we can extend φ to a ring homomorphism $A \to \mathbf{Z}/n\mathbf{Z}$ by putting $\varphi(\gamma) = \varphi(\beta)\varphi(m)^{-1}$.

Secondly, it is, for the ring A, not necessarily true that each prime ideal $\mathfrak{p}$ dividing an expression $a + b\alpha$, with a, $b \in \mathbf{Z}$ coprime, is a first degree prime ideal in the sense that $\# A/\mathfrak{p}$ is a prime number. In addition to the first degree prime ideals, one may also encounter prime ideals $\mathfrak{p}$ that contain a prime number p dividing the index $[A : \mathbf{Z}[\alpha]]$ of additive groups and that intersect $\mathbf{Z}[\alpha]$ in a first degree prime ideal of $\mathbf{Z}[\alpha]$; such p divide drs. In the rest of this section we call these prime ideals *exceptional*. In order to compute the prime ideal factorizations of the expressions $a + b\alpha$ one needs to construct the exceptional prime ideals $\mathfrak{p}$ as well as the corresponding valuations. The existence of an efficient algorithm for doing this follows from [30, Theorem 4.9, and the discussion following its proof]; often there are faster ad hoc ways to proceed than the one indicated in [30].

3.6. *Searching for prime elements.* Denote by ω_d the volume of the unit ball in $\mathbf{R}^d$. We have $\omega_d = \pi^{d/2}/\Gamma(1 + \frac{d}{2})$, where $\Gamma(1 + \frac{d}{2})$ can be calculated from $\Gamma(1 + z) = z\Gamma(z)$, $\Gamma(1) = 1$, and $\Gamma(\frac{1}{2}) = \sqrt{\pi}$. Next put $v_d = (4/d)^{d/2}/\omega_d$, which is $((2 + o(1))/(\pi e))^{d/2}$ for $d \to \infty$, and

$$C = (v_d \cdot \sqrt{\Delta} \cdot B_2)^{2/d}, \qquad M = [v_d \cdot \sqrt{\Delta}].$$

With this notation, one searches among all non-zero elements $\gamma \in A$ for which $\frac{1}{d}\sum_\sigma |\sigma\gamma|^2 \leq C$, with σ ranging over the embeddings $K \to \mathbf{C}$; if we write $\gamma = \sum_{i=0}^{d-1} q_i \alpha^i$ with $q_i \in \mathbf{Q}$, then $\frac{1}{d}\sum_\sigma |\sigma\gamma|^2 = \sum_{i=0}^{d-1} q_i^2 |t|^{2i/d}$. One of the purposes of the search is to find, for each first degree or exceptional prime ideal $\mathfrak{p}$ with $M < \mathfrak{N}\mathfrak{p} \leq B_2$, a non-zero element $\pi_\mathfrak{p} \in \mathfrak{p}$ with $|\mathbf{N}(\pi_\mathfrak{p})| \leq M\mathfrak{N}\mathfrak{p}$. One can check whether a given element γ can play the role of $\pi_\mathfrak{p}$ for some $\mathfrak{p}$ if one knows the norm of γ, as in 3.1.

It is a consequence of the Minkowski lattice point theorem that at the end of the search for each $\mathfrak{p}$ an element $\pi_\mathfrak{p}$ has been found. The ideal generated by $\pi_\mathfrak{p}$ is then not necessarily equal to $\mathfrak{p}$, but it is equal to $\mathfrak{p}$ multiplied by an ideal of norm at most M. From $\mathfrak{N}\mathfrak{p} > M$ one sees that $\mathfrak{p}$ occurs exactly once in $\pi_\mathfrak{p}$.

The theory of sphere packings implies that the choice of v_d above is not the best one, and in practice one would do wise to experiment with smaller values for v_d. For the purposes of a complexity analysis the value given above is good enough, since any feasible choice for v_d is outweighed by $\sqrt{\Delta}$.

3.7. *Factoring elements.* We shall write G' for the set of $\pi_\mathfrak{p}$'s found in 3.6, with $\mathfrak{p}$ ranging over the set of first degree or exceptional prime ideals for which $M < \mathfrak{N}\mathfrak{p} \leq B_2$. In addition, we write H for the multiplicative group of non-zero elements $\gamma \in K$ for which the fractional ideal $A\gamma$ is built up from the prime ideals of norm $\leq M$. All units of A clearly belong to H. The set G' and the group H will play the role that the set G and the group of units of $\mathbf{Z}[\alpha]$ played in Section 2. For example, the algorithm of Section 2 requires that we write certain expressions of the form $a + b\alpha$ as a product of powers of elements of G and a unit, and similarly for certain alternating products of such expressions. In the modified algorithm, we write expressions of the same type as a product of elements of G' and an element of H. To determine which power of $\pi_\mathfrak{p} \in G'$ occurs in a given expression one proceeds exactly as in Section 2, using 2.6, except if $\mathfrak{p}$ is exceptional; in the latter case one needs to apply the valuations mentioned in 3.5.

3.8. *Generators for H.* The next step is to find a multiplicative representation for the elements of H, by means of a set U' that plays the role of U. One can attempt to find such a set U' by means of the method indicated in 3.1. Namely, during the search in 3.6 one also keeps track of elements that are entirely built up from prime ideals of norm at most M. If one is lucky, one obtains in this way not only a set U of generating units, but also generators for each of the prime ideals of norm at most M, either directly or by combining a few elements that are found. The set U' then consists of U together with the generators of those prime ideals, and as in 3.1 one can modify the elements $\pi_\mathfrak{p} \in G'$ found in 3.6 so as to obtain true generators for the larger prime ideals $\mathfrak{p}$. Altogether this situation is very similar to what we had earlier, the main difference being that $\mathbf{Z}[\alpha]$ is replaced by A. In particular, A is a unique factorization domain in this case.

In the examples that we tried the above is, essentially, what happened. In general one cannot expect to be so fortunate. In the first place, the ring A need not be a unique factorization domain, in which case there do not exist generators for all primes of norm at most M. A more serious difficulty is caused by the possibility that A does not have a set of "small" generating units. If this occurs, then not only the units, but also generators of some of the prime ideals may be hard to find. It seems likely that this difficulty may actually be encountered in practice. Since we have no experience with it, we do not know which of our ideas for dealing with it is to be recommended for practical use. We shall just make a few remarks of a theoretical nature, which indicate that there do exist satisfactory ways to solve the problem. In this discussion we make no assumptions on unique factorization in A or about the units of A.

One possibility is to use an algorithm of Buchmann, see [4; 30, Theorem 5; 6, Section 6]. If properly modified and interpreted, this algorithm yields a set of independent generators U' for the group H, and these generators are such that there is a fast algorithm that given $\gamma \in H$ finds the unique expression of γ as a product of powers of elements of U'. This means that U' can in a satisfactory way play the role of U.

A second possibility is to use Theorem 6.2 of [30] in order to find a set of generators for H. In order to convert this set of generators into an independent set U' for which a fast algorithm as just indicated exists, one can apply linear algebra over $\mathbf{Z}$ (see [20]) as well as basis reduction techniques (cf. [18]).

A third possibility is not to bother about finding a generating set for H at all, but waiting for elements of H to produce themselves in the course of the algorithm. For example, every B_2-smooth expression $a + b\alpha$ that is found gives, upon division by an appropriate product of elements of G', an element of H, and likewise for the alternating products of expressions $a + b\alpha$ that arise from the partial relations. The collection of all elements of H that produce themselves in this way generates a subgroup H' of H. It is not guaranteed that H' equals H, but since we never encounter elements of H outside of H' this is of no concern to us. We do need to convert the given set of generators for H' into a set U' of independent generators for H'. This can be done as indicated above. See also Remark 5.4.

Once a suitable set U' has been found, it is necessary to find, for each $u \in U'$, an element of $\mathbf{Z}/n\mathbf{Z}$ that can meaningfully be called $\varphi(u)$. If U' is contained in A this presents no problem, since φ is defined on A. If U' is not contained in A we can proceed as we did with A itself; this can be done if n is free of prime factors $\leq M$, a condition that can easily be checked.

It is to be remarked that the third possibility for finding U' mentioned above can in principle be extended to G': do not search for elements $\pi_{\mathfrak{p}}$, but wait for them (or for elements that are just as good) to produce themselves in the course of the algorithm. This approach might be feasible for number fields for which Δ is much larger, and it might therefore be useful if one wants to apply the number field sieve to arbitrary positive integers n. The resulting algorithm is, in a somewhat different formulation, discussed in 9.1. As we shall see in 9.6, there is also a variant of the number field sieve that dispenses with G' and U' altogether.

In several of the manipulations with elements of H and G' that we just sketched it happens that the elements u that one is interested in arise as the product of powers of certain other elements and their inverses. In this case it may be laborious to calculate the explicit expression $u = \sum_{i=0}^{d-1} q_i \alpha^i$ of u in terms of the powers of α. It is good to keep in mind that this calculation can usually be avoided. This is because the information on u that the algorithm really needs—such as the vector $\nu(u)$ defined in Section 5, the argument of u under one particular embedding $K \to \mathbf{C}$, and the value of $\varphi(u) \in \mathbf{Z}/n\mathbf{Z}$—can all be derived from the given product representation for u.

4. Sieving

In this section we describe how the search for full and partial relations can be carried out. In the notation of Section 2, these relations correspond to pairs of coprime integers (a, b) such that $a + bm$ is B_1-smooth, except for at most one prime factor $< B_3$, and such that $a + b\alpha$ is B_2-smooth, except for at most one prime ideal of norm $< B_4$.

From a theoretical point of view one can solve the problem of finding these pairs (a, b) by applying the elliptic curve smoothness test [25, Section 4.3] to each individual pair, because smoothness of the algebraic integer $a + b\alpha$ is equivalent to smoothness of the integer $a^d - t(-b)^d$, cf. 2.6. In this section we present a more practical way to find the pairs (a, b). For the purposes of the run time analysis in Section 6 the two methods are equivalent.

We describe a method to find pairs (a, b) that satisfy the conditions (i), (ii), and (iii) of 2.8 for some fixed positive value of b, and for a ranging over an interval $[a_{min}, a_{max})$. This method is applied to all b in $[1, b_{max}]$ that are to be processed. There is no particular order in which this has to be done. We shall see, however, that one can gain some efficiency by processing the b's in order. This is to be kept in mind if the search for relations is carried out in parallel on many independent processors: it is better to assign a range of consecutive b-values to each processor than some arbitrary set of b-values. An entirely different way to organize the sieving step is described in [2].

The values of a_{min} and a_{max} are best determined empirically; see Sections 6 and 8 for theoretical and practical choices. It is not necessary to make a choice for b_{max}, since one can simply continue until the number of full relations plus the number of independent cycles among the partial relations is larger than $\#I$, the cardinality of the factor base. With growing b, however, the probability that both $a + bm$ and $a + b\alpha$ are smooth gets smaller, and quite noticeably so. This means that if B_1 and B_2 have been chosen too low, then one might never find sufficiently many relations. See Section 8 for examples of b_{max} and Section 6 for a theoretical estimate.

4.1. *Two sieves.* Fix some positive value for b. Testing the numbers $a + bm$ with $a \in [a_{min}, a_{max})$ for B_1-smoothness can be done by means of a sieve over a, because p divides $a + bm$ for all a that are $-bm$ mod p. After sieving with all $p \leq B_1$, one can identify the pairs (a, b) that have a reasonable chance to satisfy 2.8(ii), and one has to inspect the corresponding $a + b\alpha$'s for B_2-smoothness, cf. 2.8(iii). If there are only a few candidates, one can do this using trial division of the norms $N(a + b\alpha)$ (after checking that 2.8(i) holds), cf. 2.6. In practice it will be much faster to apply a second sieve, again over the entire interval of a-values, because the number of candidates, for a proper choice of the factor base, will be considerable. A sieve can be applied because the first degree prime ideal corresponding to a pair $(p, c$ mod $p)$, as in 2.6, occurs in $a + b\alpha$ for all a that are $-bc$ mod p.

Only pairs (a, b) for which both $a + bm$ (after the first sieve) and $a + b\alpha$ (after the second sieve) are likely to satisfy the smoothness conditions of 2.8(ii) and 2.8(iii), are subjected to further gcd and trial division tests to see if the pair indeed gives rise to a full or a partial relation. The rest of this section is devoted to a more detailed description of a possible implementation of the sieving step.

4.2. *The rational sieve.* We describe how the $a + bm$, for some fixed b, can be sieved for B_1-smoothness. For all $a \in [a_{min}, a_{max})$ initialize the sieve locations s_a as zero. Next, for all primes $p \leq B_1$, replace s_a by $s_a + \log p$ for all $a \in [a_{min}, a_{max})$ that are $-bm$ mod p.

If, after all primes $p \leq B_1$ have been processed, a sieve location s_a is close to $\log |a + bm|$, then it is quite likely that $a + bm$ is B_1-smooth. If $s_a \geq \log(a + bm) - \log B_3$, then $a + bm$ is probably B_1-smooth except for a factor that is at most B_3; if $B_3 < B_1^2$, then this factor will be prime, if we assume that $a + bm$

is square-free. The event that $s_a \geq \log(a + bm) - \log B_3$ is called a *report* (but see 4.3).

4.3. *Efficiency considerations.* Because we do not sieve with prime powers, not all smooth $a + bm$ are caught in the sieve: numbers $a + bm$ that are smooth and not square-free may be overlooked. This is only a first step in speeding up the sieving without affecting its yield by too much. In practice quite a few more smooth $a + bm$ will be missed, because 4.2 is only an idealized version of what actually happens. The s_a, for instance, are usually represented by 8-bit (1-byte) integers. Consequently, $\log p$ is rounded to the nearest integer, and the base of the logarithm is chosen large enough so that overflow is avoided when 8-bit (or 7-bit, see below) integers are added. Furthermore, one often does not sieve with the small primes below a certain small prime bound, or one replaces them by a small power. This means that one should use $\log(a + bm) - \log B_3 - B_5$ instead of $\log(a + bm) - \log B_3$ while checking for reports, for some B_5 that depends on the small prime bound. The small prime bound and the corresponding B_5 are best determined empirically.

It is often a good idea to allow negative s_a, i.e., 7-bit integers plus one bit for the sign. This makes it possible to initialize the s_a as $-\log(a+bm)+\log B_3+B_5$ so that the report-check can be replaced by a non-negativity check, which is often faster. In many architectures four consecutive 8-bit s_a's can be checked simultaneously for non-negativity by means of one 32-bit 'and'-operation with the proper mask. Since a will be small compared to bm, all s_a can be initialized to the same rounded value $-\log(bm) + \log B_3 + B_5$. This often allows a simultaneous initialization of several consecutive s_a's. All these changes are intended to decrease the cost of the sieving step, while some of them have a negative effect on the performance. Care should be taken that the cost/performance ratio does *not* increase.

The computation of $-bm \bmod p$ requires a (multi-precision) division by p, unless $-(b-1)m \bmod p$ is known, in which case a few additions and comparisons suffice. This makes consecutive processing of the b's slightly faster than processing them in random order. A similar remark applies to the second, algebraic sieve.

4.4. *The algebraic sieve and trial division.* We describe how to process the reports from 4.2 in order to locate the pairs (a, b) that satisfy conditions 2.8(i), (ii), and (iii), for the same fixed b as in 4.2. The improvements of 4.3 are taken into account.

First, check for reports: let $A = \{a : s_a \geq 0\}$ be the set of a's at which reports occur, replace s_a for $a \in A$ by some moderately small negative number B_6, and leave the other s_a unchanged. Next use a sieve to replace, for all first degree prime ideals $\mathfrak{p}$ with $\mathfrak{N}\mathfrak{p} \leq B_2$, the number s_a by $s_a + \log p$ for all $a \in [a_{\min}, a_{\max})$ with $a \equiv -bc \bmod p$, where $(p, c \bmod p)$ corresponds to $\mathfrak{p}$.

Finally, do the following for all $a \in A$ for which $s_a \geq 0$. If $\gcd(a, b) = 1$, then compute $\log |a^d - t(-b)^d|$—for which a crude floating point approximation to the integer $a^d - t(-b)^d$ suffices—and check if $s_a \geq \log |a^d - t(-b)^d| + B_6 - \log B_4$, cf. 2.6. If that turns out to be the case, attempt to factor $|a + bm|$

using trial division by the primes $\leq B_1$. If $|a + bm|$ is B_1-smooth, except for at most one prime factor $< B_3$, compute $|a^d - t(-b)^d|$, and attempt to factor this number using trial division by the primes $\leq B_2$. If $|a^d - t(-b)^d|$ turns out to be B_2-smooth, except for at most one prime factor $< B_4$, then a pair (a, b) satisfying 2.8(i), (ii), and (iii) has been found.

We introduced B_6 because $|a^d - t(-b)^d|$ can vary considerably over the a-interval. This implies that no uniform negative initialization of the s_a as in 4.3 can be used, which would allow an easy non-negativity check after the sieving. To avoid the computation of $\gcd(a, b)$ (and of $\log|a^d - t(-b)^d|$) for all $a \in A$ (with $\gcd(a, b) = 1$), we initialize the s_a for $a \in A$ as B_6, and we only compute the corresponding gcd (and possibly the logarithm) if s_a has at least made it to a non-negative number after the sieving. This saves some time, but it also introduces an extra inaccuracy, because values of a close to a zero of $a^d - t(-b)^d$ can be overlooked if B_6 is chosen too small. The value for B_6 is best determined empirically. Notice that overflow may occur in s_a for $a \notin A$, which one can avoid by changing these s_a to the smallest negative value that they can assume instead of leaving them unchanged during the report-check.

One can also compute $\gcd(a, b)$ before putting a in A. In that way fewer locations have to be checked after the second sieve. On the other hand, many more gcd's would have to be computed than in the version given above, because in that version the $a \in A$ for which $s_a < 0$ after the second sieve are cast out. It depends on the relative speed of the various operations which version is preferable.

4.5. *Remark.* If $a_{\max} - a_{\min}$, the number of memory locations needed for the sieve, is more than can be allocated, then the interval $[a_{\min}, a_{\max})$ should be partitioned into subintervals to which 4.2 and 4.4 can be applied. If the subintervals are processed in order (of the a's), then one can easily arrange an efficient transition from one subinterval to the next, by remembering the last visited a-value for each p and $\mathfrak{p}$.

4.6. *Choice of B_3 and B_4.* We conclude this section with a few remarks concerning the choice of the large prime bounds B_3 and B_4. In the theoretical analysis in Section 6 the partial relations play no role, cf. 6.2 and 6.5. This implies that the choices $B_3 = B_1$ and $B_4 = B_2$, for which all relations are full relations, are good enough from a theoretical point of view. In practice, however, partial relations make the algorithm run substantially faster, as can be seen in Section 8. The choice of B_3 and B_4 depends on various considerations. In the first place, one has to choose them such that $B_3 < B_1^2$ and $B_4 < B_2^2$ to avoid the problem of having to factor the remaining factor of $|a + bm|$ or $|a^d - t(-b)^d|$ after trial division by the primes $\leq B_1$ or $\leq B_2$, respectively. Large choices within the respective ranges result in many reports, a slow performance of the sieving step, but a very high yield. This may sound good, but large primes near the lower end of the range have a higher probability to be matched in a cycle, whereas the majority of the other partial relations will turn out to be useless, after having slowed down the sieving step and clogged up the disks. A reasonable choice for B_3 seems to be somewhere between $B_1^{1.2}$ and $B_1^{1.4}$, and similarly for B_4.

5. FINDING THE UNIT CONTRIBUTION

Let the notation be as in Sections 2 and 3, and let a, b be coprime integers for which $a + b\alpha$ is B_2-smooth. Then the ideal generated by $a + b\alpha$ can be written as the product of first degree prime ideals $\mathfrak{p}$ of norm $\leq B_2$:

$$(a + b\alpha) = \prod \mathfrak{p}^{e(\mathfrak{p})}.$$

In 2.6 we saw how the $e(\mathfrak{p}) \in \mathbf{Z}_{\geq 0}$ can be determined. The element $a + b\alpha$ itself can now be written as

$$(5.1) \qquad a + b\alpha = \prod_{u \in U} u^{e(u)} \cdot \prod_{g \in G} g^{e(g)},$$

where $e(g) = e(\mathfrak{p})$ if $g = \pi_{\mathfrak{p}}$, and with integers $e(u)$ that remain to be found.

Of course, one can find the unit $\prod_{u \in U} u^{e(u)}$ by computing $(a+b\alpha) \cdot \prod_{g \in G} g^{-e(g)}$ in the number field. Given a sufficiently large table of products of elements of U and their inverses, the $e(u)$ can then be found by table look-up. For very "small" fields this will probably work quite satisfactorily. However, when the field is a bit "larger", it will be quite slow, due to the arithmetic in the number field, which consists of fairly expensive polynomial multiplications and divisions modulo f. In this section we describe a faster method for determining the $e(u)$. The method keeps track of some extra information per generator $g \in G$, and uses vector additions instead of arithmetic in the number field.

Let $U = \{u_0, u_1, \ldots, u_l\}$ be as in Section 3. Choose l embeddings $\varphi_1, \varphi_2, \ldots, \varphi_l$ of K into $\mathbf{C}$ such that no two of the φ_i are complex conjugates. This can be done as follows. Let f have r_1 real roots $\alpha_1, \alpha_2, \ldots, \alpha_{r_1}$, and $2r_2 = d - r_1$ non-real complex roots $\alpha_{r_1+1}, \alpha_{r_1+2}, \ldots, \alpha_d$, with $\alpha_{r_1+r_2+j}$ the complex conjugate of α_{r_1+j}, for $1 \leq j \leq r_2$. In Section 3 we saw that $l = r_1 + r_2 - 1$. For $1 \leq i \leq l$ let φ_i be the embedding $K \to \mathbf{C}$ that maps $\sum_{j=0}^{d-1} q_j \alpha^j$ to $\sum_{j=0}^{d-1} q_j \alpha_i^j$.

For $x \in K$, $x \neq 0$, let $\nu(x)$ be the l-dimensional real vector with ith coordinate equal to $\log |\varphi_i(x)| - (\log |\mathbf{N}(x)|)/d$, for $1 \leq i \leq l$; if x is taken to be a unit of $\mathbf{Z}[\alpha]$ then $|\mathbf{N}(x)| = 1$ so that the term $-(\log |\mathbf{N}(x)|)/d$ can be omitted. Note that $\nu(u_0) = 0$. Let W be the $l \times l$ matrix having $\nu(u_i)$ as its ith column, for $1 \leq i \leq l$. The image of the group of units of $\mathbf{Z}[\alpha]$ under ν is a lattice in $\mathbf{R}^l$, a basis for this lattice being given by the columns of W.

To find the $e(u) \in \mathbf{Z}$ that satisfy (5.1) we notice in the first place that $v = (a + b\alpha) \cdot \prod_{g \in G} g^{-e(g)}$ is a unit, and that

$$(5.2) \qquad \nu(v) = \nu(a + b\alpha) - \sum_{g \in G} e(g) \cdot \nu(g).$$

Since v is a unit, $\nu(v)$ is in the lattice spanned by the columns of W, and more in particular $\nu(v) = W \cdot (e(u_1), e(u_2), \ldots, e(u_l))^T$, so that the $e(u_i)$ for $1 \leq i \leq l$ are the entries of the vector $W^{-1} \cdot \nu(v)$ and can be computed as such.

It remains to determine $e(u_0)$. If f has at least one real root, then we took $u_0 = -1$ in Section 3, and furthermore we specified some particular real embedding such that the other u_i and all $g \in G$ are positive under this embedding, see Remark 3.2. So, we put $e(u_0) = 0$ if $a + b\alpha$ is positive under this embedding, and $e(u_0) = 1$ otherwise. If f has no real roots, then u_0 is some root of unity. In this case, choose some particular complex embedding, and select $e(u_0)$ in such a way that the arguments (angles) of this complex embedding of the left and right hand sides of (5.1) match.

In practice the mapping ν and the entries of the matrix W^{-1} are only computed in limited precision, and the entries of the vector $W^{-1} \cdot \nu(v)$ are rounded to integers. To avoid problems with limited precision computations, it helps to select (or to change) U such that the columns of W form a reduced basis for the lattice that they span. It also helps to select (or to change) the elements $g \in G$ such that the coordinates of $W^{-1} \cdot \nu(g)$ lie between $-\frac{1}{2}$ and $\frac{1}{2}$; one can achieve this by multiplying g by an appropriate product of units (to be determined with the help of ν). In our implementation, the vectors $W^{-1} \cdot \nu(g)$, for $g \in G$, were computed once and for all and kept in a file.

5.3. *Remark.* The same method can be applied to determine the unit contribution in a cycle C, where we are dealing with $\prod_{(a,b) \in C} (a + b\alpha)^{s(a,b)}$ instead of $a + b\alpha$ in (5.1): simply replace $\nu(a + b\alpha)$ in (5.2) by $\sum_{(a,b) \in C} s(a,b) \cdot \nu(a + b\alpha)$. The case of the free relations (see 2.13) is even easier, since $\nu(p)$ is the zero vector.

5.4. *Remark.* If the set U as determined in Section 3 fails to generate the unit group, or if one decides not to bother determining U at all (cf. 3.1 and 3.8), then the algorithm described in this section needs to be modified in the sense that the units must be processed as they come along. At each stage, one needs to keep track of a set of independent generators for the group generated by the units that have appeared so far. If a unit u is encountered for which $\nu(u)$ does not belong to the lattice spanned by the images, under ν, of the current set of generators, then this set of generators needs to be updated. This can be done by means of lattice basis reduction techniques. It is likely that only a few of such updates will be necessary. In the cases that we tried difficulties of this nature did not arise. If they do arise, then the remark made at the end of Section 3 may help to minimize the arithmetic that needs to be done with the units.

6. RUN TIME ANALYSIS

In this section we present a heuristic estimate for the run time of the number field sieve. Currently there are several factoring algorithms that have a subexponential expected run time, such as the continued fraction algorithm, the quadratic sieve algorithm and its variants, the elliptic curve method, the number field sieve, Dixon's random squares algorithm, Vallée's two-thirds algorithm, and the class group relations method. Only for the last three algorithms has a rigorous analysis of the expected run time been given [31; 35; 41]. For the other algorithms the only available run time analysis is based on heuristic estimates, but in practice they perform better than the rigorously analyzed ones.

Each of the algorithms mentioned generates, implicitly or explicitly, a sequence of integers of which only the smooth ones are useful. Depending on the algorithm and on its implementation, these integers are constructed deterministically or drawn from a certain distribution. In all cases, the expected number of smooth integers in the sequence plays an important role in the run time analysis. A satisfactory estimate for this expected number can be given if the integers are independently drawn from the uniform distribution on the interval $[1, B]$, for some upper bound B. However, none of the algorithms that we mentioned satisfies this condition. To obtain a heuristic analysis, one simply *assumes* that the smoothness probabilities are the same as in the independent, uniform case. Only for the random squares algorithm, the two-thirds algorithm, and the class group relations method has this actually been *proved*, and this leads to a rigorous analysis of their expected run times.

For the other algorithms, including the algorithm described in this paper, nothing better can presently be given than a heuristic analysis. This is not fully satisfactory, but it is better than having nothing at all. Such heuristic analyses add to our understanding of algorithms that are practically useful. They enable us to make comparisons between different algorithms, and to make predictions about their practical performance. If one insists on having fully proved theorems, then the best one can currently do is explicitly formulating all heuristic assumptions that enter into the analysis. For examples of such theorems we refer to [34]. For one factoring algorithm, the random class groups method, one of these heuristic assumptions turned out to be incorrect, and consequently the heuristic subexponential run time estimate for that algorithm had to be withdrawn (see [31]).

For the number field sieve the heuristic run time analysis is unusually laborious, and it is carried out in some detail in [7]. The algorithm in the present paper is sufficiently similar to that in [7] that we may content ourselves with indicating how the analysis in [7] needs to be modified, and what the outcome of the modified analysis is.

Our estimates will depend on $N = r^e - s$ rather than on the divisor n of N; in most cases N will be not much larger than n. We use the notation $L_x[v, \lambda]$ introduced in the Introduction. Also, the expression $L_N[v, \lambda + o(1)]$ will be abbreviated to $L_N[v, \lambda]$, here the $o(1)$ is for $N \to \infty$, uniformly for r, s in a finite set. We shall express our final estimates in the latter notation. We note that this makes sense only if r, s are fixed, or range over a finite set, and e tends to infinity.

6.1. *Probability of smoothness.* The result that makes the L-function useful in estimating smoothness probabilities reads as follows (cf. [25, (3.16); 7, Section 10]). Let $C \subset \mathbf{R}^4$ be a compact set such that for all $(\lambda, \mu, w, v) \in C$ one has $\lambda > 0$, $\mu > 0$, and $0 < w < v \le 1$. Then the probability that a random positive integer $\le L_x[v, \lambda]$ is $L_x[w, \mu]$-smooth equals $L_x[v - w, -\lambda(v - w)/\mu + o(1)]$ for $x \to \infty$, uniformly for (λ, μ, w, v) in C.

6.2. *Parameter choice as a function of the degree.* We begin by indicating the optimal choices of the parameters as a function of N and the degree d of the number field. These are derived from 6.1 by means of the heuristic argument that was presented in support of Conjecture 11.4 in [7]. The main change that needs to be made is that the upper bound for $|(a + bm)\mathbf{N}(a + b\alpha)|$ used in [7] is replaced by the smaller value $(a_{\max} + b_{\max}m)(a_{\max}^d + b_{\max}^d|t|)$, where $m \approx N^{1/d}$ and where we assume $a_{\min} = -a_{\max}$. Following this change through the entire argument one finds that optimal choices of the parameters are obtained if all of $a_{\max}, b_{\max}, B_1$, and B_2 are taken equal to

$$\exp\left(\left(\tfrac{1}{2} + o(1)\right)\left(d\log d + \sqrt{(d\log d)^2 + 2\log(N^{1/d})\log\log(N^{1/d})}\right)\right),$$

the $o(1)$ being for $e \to \infty$, uniform for bounded r and s and for d in the region $1 < d^{2d^2} < N$. (The analysis in [7] assumes that $a_{\max} = b_{\max}$, $B_1 = B_2$, but this makes no difference.) In addition, we take $B_3 = B_1$ and $B_4 = B_2$, so that only full relations are considered (see 6.5). The size of the factor base and the number of full relations that one expects to find are given by the same expression. The typical size of the numbers $|(a + bm)\mathbf{N}(a + b\alpha)|$ that one wants to be smooth is

$$\exp\left(\left(\tfrac{1}{2} + o(1)\right)\left(d^2\log d + 2\log(N^{1/d}) + d\sqrt{(d\log d)^2 + 2\log(N^{1/d})\log\log(N^{1/d})}\right)\right).$$

The run time for the sieving in Step 2 and for the solution of the linear system in Step 3 each come out to be

$$\exp\left((1 + o(1))\left(d\log d + \sqrt{(d\log d)^2 + 2\log(N^{1/d})\log\log(N^{1/d})}\right)\right).$$

The other parts of the algorithm take less time, with the possible exception of the search for G and U in Step 1, since this search has no equivalent in the algorithm of [7]; this point is discussed in 6.4.

6.3. *Optimal parameters.* The optimal choice for d as a function of N is given by

$$d = \left(\frac{(3 + o(1))\log N}{2\log\log N}\right)^{1/3} \qquad \text{for } e \to \infty$$

uniformly for r, s in a finite set. With this choice for d, the choices for $a_{\max}$, $b_{\max}$, B_1, B_2, B_3, and B_4 made in 6.2 are $L_N[\tfrac{1}{3}, (2/3)^{2/3}]$. The typical size of the numbers $|a + bm|$ and $|\mathbf{N}(a + b\alpha)|$ is $L_N[\tfrac{2}{3}, (2/3)^{1/3}]$, so the numbers $|(a + bm)\mathbf{N}(a + b\alpha)|$ that one wants to be smooth are about $L_N[\tfrac{2}{3}, (16/3)^{1/3}]$; this is $N^{o(1)}$, as announced in the introduction. The run time of the entire algorithm, with the possible exclusion of the search in the number field in Step 1, is $L_N[\tfrac{1}{3}, (32/9)^{1/3}]$.

6.4. *Complexity of the search in the number field.* As we saw in Section 3, the search for U and G described in the first half of Section 3 is not likely to work in all cases. For this reason we consider instead the modifications described in 3.6 and 3.8. A routine calculation shows that the determination of G' in 3.6 can, for the parameter choices in 6.2 and 6.3, be performed within the same time limit. For the methods to determine U' that were indicated in 3.8 this is not so clear. To illustrate the difficulty, let us consider the algorithm of Buchmann that was mentioned in 3.8. Its run time is, according to [30, Theorem 5; 6, Section 6], bounded by $(\log \Delta)^{cd} \cdot \sqrt{\Delta}$ for some absolute constant c. In our case we have $\Delta = d^{(1+o(1))d}$ for $e \to \infty$ (uniformly for bounded r, s), so that the run time estimate of Buchmann's algorithm becomes $d^{(c+\frac{1}{2}+o(1))d}$. Since the run time in 6.3 is $d^{(4+o(1))d}$, this leads to the question whether one can take $c \leq 3\frac{1}{2}$. We do not know the answer to this question, but we consider it likely that at least one of the methods suggested in 3.8 will run in time at most $d^{(4+o(1))d}$ with our choice of parameters. If this is the case, then the run time $L_N[\frac{1}{3}, (32/9)^{1/3}]$ mentioned in 6.3 indeed applies to the entire number field sieve. If it is not the case, we can still claim this run time for the algorithm of [7], when applied to integers of our special form. In the examples that we did, the search for U and G took only a very small fraction of the total run time.

It may also be possible to justify, along similar lines, the run time given in 6.2 as a function of d, though perhaps not for as wide a range of d as indicated in 6.2.

6.5. *Remark.* Because of our choice $B_3 = B_1$ and $B_4 = B_2$, partial relations and cycles among them were not considered in the version of the algorithm analyzed above. The use of partial relation is important for the practical performance of the number field sieve, as we shall illustrate in Section 8. Nevertheless, it is unlikely that the use of partial relations will affect the run time estimate by more than a factor $L_N[\frac{1}{3}, 0]$. For a run time analysis of the cycle finding algorithm and a discussion of the expected number of cycles we refer to [28].

7. ADDITIONAL REMARKS

7.1. *Using more number fields.* Instead of using a single number field, as in 2.5, one can consider using several fields. Because the probability of finding relations decreases with growing b, this might be advantageous, because for each number field one can start afresh with the small b values. If we use smoothness bounds B_{1i} and B_{2i} for the ith number field K_i, then we need approximately $\max_i\{\pi(B_{1i})\} + \sum_i(\#U_i + \#G_i)$ relations, where U_i generates the units and G_i the first degree prime ideals of norm $\leq B_{2i}$ in K_i. Hence no K_i should be used that contributes fewer than $\#U_i + \#G_i$ relations.

To give an example of this multi-field approach, suppose that we want to factor an integer n of the form $2^{8e} + 1$, with e a positive integer, using number fields of degree 4. Direct application of the construction in 2.5 leads to the field $\mathbf{Q}(\zeta^2)$, where ζ is a primitive 16th root of unity; ζ^2 is a zero of the polynomial $X^4 + 1$, and it maps to $2^{2e} \bmod n$ under φ. Two other fourth degree fields that can be used are the fields $\mathbf{Q}(\zeta \pm \zeta^{-1})$, where $\zeta \pm \zeta^{-1}$ satisfies the polynomial $X^4 \mp 4X^2 + 2$ and is mapped to $(2^e \pm 2^{-e}) \bmod n = (2^e \mp 2^{7e}) \bmod n$ under φ. To the free relations from 2.13 one can then add the multiplicative relations that exist between elements of different fields K_i. As can be seen in [26], we did not use this approach for $e = 64$, and as far as we know the practical importance of the multi-field approach is still unexplored. From a theoretical point of view the idea has proved to be worthwhile, see [11].

7.2. *Postponing the construction of cycles.* The construction of cycles among the partial relations to obtain relations among the a_i can be postponed until Step 3, as mentioned in 2.12. Here we sketch how this can be achieved. Given a collection of partial relations, let P_1 and P_2, as in 2.12, be the sets of large primes and large prime ideals occurring in the partial relations, and let $\bar{I} = (I \backslash U) \cup P_1 \cup P_2$. Each partial relation can be regarded as an element $\bar{v} = (\bar{v}_i)_{i \in \bar{I}}$ of $\mathbf{Z}^{\bar{I}}$. For $i \in I \backslash U$ we have $\bar{v}_i = v_i$ as in Section 2, and for each $\bar{v}$ at least 1 and at most 2 of the $\bar{v}_i$ with $i \in P_1 \cup P_2$ are non-zero. Let $\bar{V}$ be the collection of $\bar{v}$'s, and let F be the collection of full relations that have been found.

Given $\bar{V}$, we attempt to find more than $\#\bar{I} - \#F$ linearly independent linear combinations among its elements for which the entries corresponding to the $i \in P_1 \cup P_2$ are even. One way to do this is by means of the methods from [22; 37]. It is easy to see that such linear combinations correspond to cycles among the partial relations. With appropriate signs, they can be turned into relations among the a_i (as in (2.3)), where the unit contribution can be determined as before. Combined with the full relations this gives more than $\#\bar{I}$ relations among the a_i, so that Step 3 can be completed in the usual manner.

There is no obvious way to see if a collection of partial relations will indeed give rise to more than $\#\bar{I} - \#F$ linear combinations as above. In practice one could first use the cycle counter from [28], and only proceed with the matrix step above if there are enough cycles.

7.3. *Double large primes.* Following the approach from [28], we can allow two large primes exceeding B_1 in $a + bm$, or two large prime ideals of norms larger than B_2 in $a + b\alpha$. This variant of the algorithm turned out to be much slower than the version described in Section 2. This was caused by the dramatic growth of the number of reports and trial divisions in the sieving step. Most of these trial divisions were fruitless, partly because the large factors remaining after trial division were often found to be prime instead of the product of two large primes. It is possible, however, that this variant becomes preferable for larger values of n than we tried.

8. EXAMPLES

The first factorization obtained by means of the number field sieve was the factorization of the 39 digit number $F_7 = 2^{2^7} + 1$, which was in fact already known (see [32]). This factorization was carried out by the fourth author in 20 hours on a Philips P2012, an 8-bit computer with 64K of memory and two 640K disk drives. With $f = X^3 + 2$ and a factor base consisting of 500 rational primes, the units -1 and $1 + (-2)^{1/3}$, and 497 algebraic primes, it took 2000 values of b and per b the integers a with $|a| \leq 4800$ to find 538 ff's and 1133 pf's with $p_2 = 1$ and $B_3 = 10000$; no fp's or pp's were used. This led to 399 cycles, which combined with the 81 free relations (cf. 2.13) sufficed to factor F_7:

$$2^{128} + 1 = 59\,64958\,91274\,97217 \cdot 57\,04689\,20068\,51290\,54721.$$

Several steps of this first number field sieve factorization were not carried out as described in the previous sections. For instance, only the numbers $a + bm$

were sieved, prime powers were included in the sieving, and for the reports both $a + bm$ and $\mathbf{N}(a + b\alpha)$ were tested for smoothness by trial division. The unit contribution was found by means of a table containing u_1^i for $|i| \leq 8$. The fourth author was able to reduce the time needed for factoring F_7 by a factor of two by using some of the methods described in Sections 2, 4, and 5. Other numbers factored by the fourth author are $2^{144} - 3$ (44 digits, in 47 hours) and $2^{153} + 3$ (47 digits, in 61 hours):

$$2^{144} - 3 = 49\,27299\,91333 \cdot 45\,25956\,52604\,77899\,16201\,09802\,72761,$$
$$2^{153} + 3 = 5 \cdot 11 \cdot 6\,00696\,43200\,64900\,87537 \cdot 3455\,98297\,79603\,41893\,82757.$$

Other, and more general, factoring methods should actually be preferred for the factorization of integers in this range. We do not know for what size of numbers the number field sieve may be expected to be faster than other, asymptotically slower methods. We do know that for numbers of the right form that have more than 100 decimal digits the number field sieve is faster than the multiple polynomial quadratic sieve method. Until the appearance of the number field sieve, the quadratic sieve was the only algorithm by which numbers in the 100+ digit range without small factors could be factored, and it still has the advantage over the number field sieve that it applies to all numbers indiscriminately.

For our number field sieve implementation at Digital Equipment Corporation's Systems Research Center we followed the same approach as for our implementations of the elliptic curve method and quadratic sieve as described in [27]. In short, this means that one central processor distributes tasks among several hundred CVAX processors, the clients, and collects their results. For a more general set-up of the number field sieve, which also allows external sites to contribute to the factoring process by means of electronic mail, we refer to [26] and also [27]. This parallelization and distribution of tasks was used only for the second step of the algorithm, the collection of relations.

For the number field sieve tasks consist of short, non-overlapping intervals of b-values. When a client is given an interval $[b_1, b_1 + 1, \ldots, b_2]$, he starts sieving all pairs a, b for $b = b_1$, $b_1 + 1$, $\ldots$, b_2 in succession, and per b for $|a|$ less than some predetermined bound. After each b, the client reports the full and partial relations that it found for that b to the central processor (possibly no relations at all), and it reports that it just processed that particular value of b. The central processor keeps track of the relations it received and the b's that have been processed. It also notices if a client dies or becomes unavailable, which occurs for instance if a workstation is claimed by its owner. In that case the b's that are left unfinished by that client can be redistributed. In this way, all positive b's will be processed, without gaps, until sufficiently many relations have been collected.

This is a more conservative approach than we use for our elliptic curve and quadratic sieve implementations. For the latter algorithms we can afford not to worry about inputs that have been distributed but that are never processed. For the number field sieve the smaller b's are noticeably better than the larger ones, so that we decided to be careful and not to waste any of them.

TABLE 1. *Four factorizations obtained with the number field sieve*

$3^{239} - 1 = 2 \cdot 479 \cdot 17209 \cdot 4\,33019\,64055\,63553\,33339\,45745\,53310\,61280$
        $44213 \cdot p67;$

$2^{373} + 1 = 3 \cdot 60427 \cdot 694\,57949\,73168\,94264\,42566\,12436\,59806\,37197\,21883$
        $18857 \cdot p60;$

$7^{149} + 1 = 8 \cdot 10133 \cdot 4\,73384\,33355\,18992\,92791\,10650\,93183\,78061\,19829\,00857$
        $39285\,01623 \cdot p66;$

$2^{457} + 1 = 3 \cdot 6885\,35756\,02053\,19573\,06063\,38968\,00918\,44825\,49047\,29193 \cdot$
        $p89.$

In Table 1 we list the first four factorizations that were obtained with our implementation of the number field sieve, with pi denoting a prime of i decimal digits. Additional data concerning these factorizations are found in Table 2. In the first case $\mathbf{Z}[\alpha]$ is a unique factorization domain. In the other three cases this is not true, but we could use instead the ring of integers of $\mathbf{Q}(\alpha)$, which does have unique factorization. This ring of integers is equal to $\mathbf{Z}[\alpha^3/2]$, $\mathbf{Z}[\alpha]+\mathbf{Z}\cdot(\alpha+2)^4/5$, and $\mathbf{Z}[\alpha^2/2]$ in the three respective cases.

Although the theoretical analysis indicates that the choice $B_1 = B_2$ is asymptotically optimal, one can imagine that in practice there are cases in which it is better to take B_1 much smaller or much larger than B_2. We have no experience with this. Introducing several fields as in 7.1 leads to an asymmetry between B_1 and B_2, see for example [11].

The first two factorizations could have been obtained with much smaller factor bases if we had used the pp's, as we did for the other two. The first entry is the first number we collected relations for; even with our restricted use of the partial relations the factor base was chosen much too large. For the last two entries our choice of factor base size turned out to be much better. This was, in particular for the third entry, more or less a matter of luck, as we had no way to guess how many partial relations would be needed to produce a given number of cycles. The experience gained with these and other numbers (see [2; 26; 28]) enables us to select the bounds B_1 and B_2 in a slightly less uncertain manner.

Before one invests a lot of computing time in the search for relations, it is wise to check if the chosen values for B_1 and B_2 are likely to work. By processing several reasonably distanced intervals of consecutive b-values, one can get a fairly cheap and accurate estimate of the total yield of full and partial relations, which should help to decide if the choices are realistic. In our later experiments we tried to select B_1, B_2, and $b_{\max}$ such that the run time of Step 2 is minimized, and such that one quarter of the final set of relations consists of full relations and the remaining three quarters are expected to be produced by cycles among the partials. This is probably rather conservative, but given how the number of cycles varies, it seems to be a safe choice; in any case, we never had to start all over again with larger bounds.

For the factorizations reported in Table 1 the first step, the determination

TABLE 2. *Data on the four factorizations*

n is factor of	$3^{239} - 1$	$2^{373} + 1$	$7^{149} + 1$	$2^{457} + 1$
# digits of n	107	108	122	138
f	$X^5 - 3$	$X^5 + 4$	$X^5 + 7$	$X^5 + 8$
m	3^{48}	2^{75}	7^{30}	2^{92}
$B_1 = B_2$	479910	287120	287120	479910
#P	40000	25000	25000	40000
#U + #G	3 + 40067	3 + 25010	3 + 24880	3 + 40012
factor base size	80070	50013	49883	80015
$B_3 = B_4$	10^8	10^8	10^8	10^8
$a_{max} = -a_{min}$	$5 \cdot 10^6$	$5 \cdot 10^6$	$5 \cdot 10^6$	$5 \cdot 10^6$
b_{max}	120000	200000	1136000	2650000
# free's	2014	1248	1222	2003
# fulls	≈ 30000	≈ 20000	10688	17625
# partials	not kept	not kept	1358719	1741365
# pf, pf pairs	≈ 25000	≈ 15000	5341	not counted
# fp, fp pairs	≈ 25000	≈ 15000	5058	not counted
# cycles with pp's	not used	not used	≈ 28000	not counted
total # cycles	> 50000	> 30000	≈ 38400	62842
run time Step 2	2 days	3 days	2 weeks	7 weeks
run time Step 3	2 weeks	4 days	5 days	2 weeks
# digits of factors	41 & 67	48 & 60	56 & 66	49 & 89

of sets of generators, turned out to be quite easy. In all four cases the set U consisted of -1 and two units of infinite order, which were not hard to come by. Determination of G by means of the method described in Section 3 never took more than fifteen minutes on a CVAX processor. In Step 2, we partitioned the a-interval into subintervals of length 500000.

One can find the cycles of length two in a trivial manner by sorting the pf's (and fp's) according to p_1 (and p_2), which is all we did to generate the cycles for the first two factorizations. For the third entry the yield was already getting quite low for b around 1100000, and we would never have been able to factor the third and the fourth number had we not used cycles involving pp's as well.

To place the run times in perspective one should keep in mind that Step 2 was performed on a network of several hundred CVAX processors, whereas Step 3 was done on a single workstation containing six CVAX processors by means of a fairly elementary Gaussian elimination program. Since these factorizations were carried out we made substantial improvements in our implementation of the third step, see [26; 28]. Other numbers we factored are a composite 115 digit factor of $3^{241} - 1$ into a p52 and a p57, a composite 108 digit factor of $6^{149} - 1$ into p36 · p79, and a composite 117 digit factor of $3^{251} - 1$ into p37 · p80. These factorizations did not produce new insights, and they were reported in the updates to [3]. Furthermore we factored the composite 148 digit factor of the ninth Fermat number, as reported in [26]. For more numbers we refer to [2].

9. GENERALIZATION

Following an idea of Buhler and Pomerance, we can attempt to generalize the number field sieve to integers n that do not have a small multiple of the form $r^e - s$, for small r and $|s|$, as follows. Select a positive integer d, an integer m that is a little smaller than $n^{1/d}$, and put $f = \sum_{i=0}^{d} f_i X^i$, where $n = \sum_{i=0}^{d} f_i m^i$ with $0 \leq f_i < m$. The algebraic number field is then defined as $K = \mathbf{Q}(\alpha)$ with $f(\alpha) = 0$, and the map $\varphi \colon \mathbf{Z}[\alpha] \to \mathbf{Z}/n\mathbf{Z}$ sends α to $(m \bmod n)$.

If one is very lucky one hits upon a value of m for which the resulting number field has a very small discriminant. This occurs, for example, if the digits f_i of n in base m are very small. In that case the algorithm as described in this paper can be applied without major changes. It is much more likely, however, that one is not so lucky, and then Steps 1 and 3 will run into serious trouble. It is debatable how probable it is that $\mathbf{Z}[\alpha]$ (or the ring of integers of K) is a unique factorization domain (see [9]); but even if it is, it is completely unrealistic to expect that the search methods discussed in Section 3 can be used to find generators for the unit group and for the first degree prime ideals. This is because the values for M and C would have to be taken prohibitively large. Standard estimates suggest that the coefficients of the elements of U and G, when written as explicit polynomials in α, are so large that they cannot even be written down in a reasonable amount of time, let alone calculated. This means that the elements of U and G must be represented in a different way, or that their computation must be avoided altogether. We discuss a variant of the number field sieve that accomplishes the latter.

9.1. *Elimination over* $\mathbf{Z}$. To describe this variant, we make the simplifying assumption that $\mathbf{Z}[\alpha]$ is the ring of integers of K; this assumption is discussed in 9.4. Also, we consider, for simplicity, only *full* relations. The sieving step provides us with many pairs of coprime integers a, b with the property that both $a + bm$ and $a + b\alpha$ are smooth:

$$a + bm = \prod_p p^{e_{a,b}(p)}, \qquad (a + b\alpha) = \prod_{\mathfrak{p}} \mathfrak{p}^{e_{a,b}(\mathfrak{p})}.$$

Here p ranges over the prime numbers $\leq B_1$, and $\mathfrak{p}$ over the first degree prime ideals of $\mathbf{Z}[\alpha]$ of norm $\leq B_2$; furthermore, the $e_{a,b}(p)$ and $e_{a,b}(\mathfrak{p})$ are non-negative integers that one can compute. Note that we use only the *ideal* factorization of $a + b\alpha$, not a factorization in terms of sets U and G. Next one looks for solutions to the system

$$\sum_{a,b} e_{a,b}(p) z_{a,b} \equiv 0 \bmod 2, \qquad \sum_{a,b} e_{a,b}(\mathfrak{p}) z_{a,b} = 0, \qquad z_{a,b} \in \mathbf{Z},$$

the sums ranging over the pairs a, b that have been found. This amounts to solving a large and sparse system of linear equations over $\mathbf{Z}$. If r_1, r_2 are as in Section 3, then one needs a little more than $r_1 + r_2$ solutions $z = (z_{a,b})$; they should be independent in a suitable sense. For each solution z, the integer

$$\prod_{a,b} (a + bm)^{z_{a,b}}$$

is the square of an integer that can be explicitly written down as a product of prime numbers $p \leq B_1$. Also, the exponents occurring in the prime ideal factorization of the algebraic number

$$(9.2) \qquad\qquad u = \prod_{a,b} (a + b\alpha)^{z_{a,b}}$$

are all equal to 0, so u is a unit. Therefore each z gives rise to a relation of the type

$$(9.3) \qquad\qquad \left(\prod_p \varphi(p)^{w(p)} \right)^2 = \varphi(u)$$

in $\mathbf{Z}/n\mathbf{Z}$, where u is a unit given by (9.2) and the $w(p)$ are integers. If one has sufficiently many such relations, then the units u become multiplicatively dependent, and one can find an explicit dependence relation by combining the techniques of Section 5 with lattice basis reduction. For this one needs to know the logarithms of the images of the units u under the embeddings $K \to \mathbf{C}$, and these can be computed from (9.2). Taking the corresponding product of the relations (9.3) one finds a relation of the type

$$\left(\prod_p \varphi(p)^{y(p)} \right)^2 = \varphi(1),$$

so that $x = \prod_p p^{y(p)}$ is a solution to (2.2), as required.

9.4. *The ring of integers.* In 9.1 we made the assumption that $\mathbf{Z}[\alpha]$ is the ring of integers of K. If this condition is not satisfied then some of the elements u produced by the algorithm may not be units. In that case the vectors $\nu(u)$ from Section 5 will not necessarily belong to a lattice, so that lattice basis reduction techniques cannot be used to find relations with integer coefficients between these vectors.

There are several ways to deal with this problem. The first is based on the following conjecture, which is quite possibly provable with present-day techniques: let d be an integer, $d \geq 2$; then for a "random" polynomial $f = \sum_{i=0}^{d} f_i X^i$ with $f_i \in \mathbf{Z}$, $f_d = 1$, the condition that $\mathbf{Z}[\alpha]$ is the ring of integers of K is satisfied with probability equal to $6/\pi^2$. This conjecture suggests that when one tries a few values for m one soon runs into one for which the condition is satisfied, so that the algorithm does not encounter the difficulty just indicated.

Alternatively, one might deal with the problem by replacing $\mathbf{Z}[\alpha]$ by the ring of integers A of K. With from some minor adjustments to the algorithm (see 3.5) this ring can take the place of $\mathbf{Z}[\alpha]$. One might object that the only known algorithms for determining A that are sufficiently fast for our purpose may fail (see [30, Section 4; 5]); more precisely, they do produce a subring A' of A, but A' may be different from A if the discriminant of f has a very large but unknown repeated prime factor. Fortunately, one can prove that A' works just as well as A if any such prime factor exceeds B_2.

9.5. *Complexity.* In the rest of this section we abbreviate $L_n[v, \lambda + o(1)]$, for $n \to \infty$, to $L_n[v, \lambda]$. Arguments similar to those in Section 6 suggest that the variant of the number field sieve that we just discussed factors any integer n in time $L_n[\frac{1}{3}, 9^{1/3}]$, where $9^{1/3} \doteq 2.0801$. A bottleneck is formed by the large linear system, which is to be solved over $\mathbf{Z}$ rather than over $\mathbf{Z}/2\mathbf{Z}$. We did invent a fairly complicated technique by which we could reduce the size of this system to approximately its square root, while preserving the sparsity; this technique depends on the availability—from a suitably modified Step 2—of many pairs of coprime integers a, b for which $a + b\alpha$ is B_2-smooth (with no condition on $a + bm$). This reduces the conjectural run time of the number field sieve to $L_n[\frac{1}{3}, (64/9)^{1/3}]$, where $(64/9)^{1/3} \doteq 1.9230$.

9.6. *Quadratic characters.* Although the ideas exposed above may have some use in practice, our discussion has been rather sketchy. This is because there exists a method that achieves the same conjectural run time $L_n[\frac{1}{3}, (64/9)^{1/3}]$ in a conceptually much simpler way. It employs quadratic characters in the number field. They were suggested by Adleman [1] as a tool to avoid both the assumption that the ring of integers of K is a unique factorization domain and the determination of the sets U and G. In [7] it was shown that quadratic characters can also be used to avoid the need to determine the ring of integers of K (cf. 9.4), a problem that is not dealt with in [1]. For a description of the method we refer to [1; 7].

It is an essential feature of the use of quadratic characters that it produces a square in the number field without producing its square root. This leads to the problem of computing square roots of very large algebraic integers. The known methods for doing this, which are discussed on [7, Section 9], lead to arithmetic operations with integers whose number of bits is roughly proportional to the square root of the run time of the entire factoring algorithm! A method proposed by Couveignes (see [15; 2]) works with much smaller numbers; it works only if the degree d is odd.

It is as yet unknown which of 9.1 and 9.6 should be preferred in practice. The first method has the disadvantage of a considerably more complicated elimination step, the second method requires substantial computations in the number field, but, as shown in [2], works quite satisfactorily if the extension degree is odd.

Coppersmith [11] showed that one can reduce the conjectural run time to $L_n[\frac{1}{3}, c]$, where $c = \frac{1}{3}(92 + 26\sqrt{13})^{1/3} \doteq 1.9019$, by using several number fields. There is no indication that the modification proposed by Coppersmith has any practical value.

REFERENCES

1. L. M. Adleman, *Factoring numbers using singular integers*, Proc. 23rd Annual ACM Symp. on Theory of Computing (STOC), New Orleans, May 6–8, 1991, 64–71.
2. D. J. Bernstein, A. K. Lenstra, *A general number field sieve implementation*, this volume, pp. 103–126.
3. J. Brillhart, D. H. Lehmer, J. L. Selfridge, B. Tuckerman, S. S. Wagstaff, Jr., *Factorizations of $b^n \pm 1$, $b = 2, 3, 5, 6, 7, 10, 11, 12$ up to high powers*, second edition, Contemp. Math.

22, Amer. Math. Soc., Providence, 1988.

4. J. Buchmann, *Complexity of algorithms in algebraic number theory*, in: R. A. Mollin (ed.), *Proceedings of the first conference of the Canadian Number Theory Association*, De Gruyter, Berlin, 1990, 37–53.

5. J. Buchmann, H. W. Lenstra, Jr., *Approximating rings of integers in number fields*, in preparation.

6. J. Buchmann, V. Shoup, *Constructing nonresidues in finite fields and the extended Riemann hypothesis*, in preparation. Extended abstract: Proc. 23rd Annual ACM Symp. on Theory of Computing (STOC), New Orleans, May 6–8, 1991, 72–79.

7. J. P. Buhler, H. W. Lenstra, Jr., C. Pomerance, *Factoring integers with the number field sieve*, this volume, pp. 50–94.

8. H. Cohen, *A course in computational algebraic number theory*, Springer-Verlag, to appear.

9. H. Cohen, H. W. Lenstra, Jr., *Heuristics on class groups*, pp. 33–62 in: H. Jager (ed.), *Number theory, Noordwijkerhout 1983*, Lecture Notes in Math. **1068**, Springer-Verlag, Heidelberg.

10. D. Coppersmith, *Fast evaluations of logarithms in fields of characteristic 2*, IEEE Trans. Inform. Theory **30** (1984), 587–594.

11. D. Coppersmith, *Modifications to the number field sieve*, J. Cryptology, to appear; IBM Research Report RC 16264, 1990.

12. D. Coppersmith, *Solving linear equations over GF(2): block Lanczos algorithm*, Linear Algebra Appl., to appear; IBM Research Report RC 16997, 1991.

13. D. Coppersmith, *Solving linear equations over GF(2) II: block Wiedemann algorithm*, Math. Comp., to appear; IBM Research Report RC 17293, 1991.

14. D. Coppersmith, A. M. Odlyzko, R. Schroeppel, *Discrete logarithms in GF(p)*, Algorithmica **1** (1986), 1–15.

15. J.-M. Couveignes, *Computing a square root for the number field sieve*, this volume, pp. 95–102.

16. J. D. Dixon, *Asymptotically fast factorization of integers*, Math. Comp. **36** (1981), 255–260.

17. T. ElGamal, *A subexponential-time algorithm for computing discrete logarithms over GF(p²)*, IEEE Trans. Inform. Theory **31** (1985), 473–481.

18. D. M. Gordon, *Discrete logarithms in GF(p) using the number field sieve*, SIAM J. Discrete Math. **6** (1993), 124–138.

19. D. M. Gordon, K. S. McCurley, *Massively parallel computation of discrete logarithms*, Advances in cryptology, Crypto '92, to appear.

20. J. L. Hafner, K. S. McCurley, *Asymptotically fast triangularization of matrices over rings*, SIAM J. Comput. **20** (1991), 1068–1083.

21. D. E. Knuth, *The art of computer programming*, volume 2, *Seminumerical algorithms*, second edition, Addison-Wesley, Reading, Massachusetts, 1981.

22. B. A. LaMacchia, A. M. Odlyzko, *Solving large sparse systems over finite fields*, Advances in cryptology, Crypto '90, Lecture Notes in Comput. Sci. **537** (1991), 99–129.

23. B. A. LaMacchia, A. M. Odlyzko, *Computation of discrete logarithms in prime fields*, Designs, Codes and Cryptography **1** (1991), 47–62.

24. S. Lang, *Algebra*, third edition, Addison-Wesley, Reading, Massachusetts, 1993.

25. A. K. Lenstra, H. W. Lenstra, Jr., *Algorithms in number theory*, Chapter 12 in: J. van Leeuwen (ed.), *Handbook of theoretical computer science*, Volume A, *Algorithms and complexity*, Elsevier, Amsterdam, 1990.

26. A. K. Lenstra, H. W. Lenstra, Jr., M. S. Manasse, J. M. Pollard, *The factorization of the ninth Fermat number*, Math. Comp. **61** (1993), to appear.

27. A. K. Lenstra, M. S. Manasse, *Factoring by electronic mail*, Advances in cryptology, Eurocrypt '89, Lecture Notes in Comput. Sci. **434** (1990), 355–371.

28. A. K. Lenstra, M. S. Manasse, *Factoring with two large primes*, Math. Comp., to appear.

29. H. W. Lenstra, Jr., *Factoring integers with elliptic curves*, Ann. of Math. **126** (1987), 649–673.

30. H. W. Lenstra, Jr., *Algorithms in algebraic number theory*, Bull. Amer. Math. Soc. **26** (1992), 211–244.

31. H. W. Lenstra, Jr., C. Pomerance, *A rigorous time bound for factoring integers*, J. Amer. Math. Soc. **5** (1992), 483–516.

32. M. A. Morrison, J. Brillhart, *A method of factoring and the factorization of F_7*, Math. Comp. **29** (1975), 183–205.

33. M. Pohst, H. Zassenhaus, *Algorithmic algebraic number theory*, Cambridge University Press, Cambridge, 1989.

34. C. Pomerance, *Analysis and comparison of some integer factoring algorithms*, pp. 89–139 in: H. W. Lenstra, Jr., R. Tijdeman (eds), *Computational methods in number theory*, Math. Centre Tracts **154/155**, Mathematisch Centrum, Amsterdam, 1983.

35. C. Pomerance, *Fast, rigorous factorization and discrete logarithm algorithms*, in: D. S. Johnson, T. Nishizeki, A. Nozaki, H. S. Wilf (eds), *Discrete algorithms and complexity*, Academic Press, Orlando, 1987, 119–143.

36. C. Pomerance (ed.), *Cryptology and computational number theory*, Proc. Sympos. Appl. Math. **42**, Amer. Math. Soc., Providence, 1990.

37. C. Pomerance, J. W. Smith, *Reduction of huge, sparse matrices over finite fields via created catastrophes*, Experiment. Math. **1** (1992), 89–94.

38. C. P. Schnorr, *Refined analysis and improvements on some factoring algorithms*, J. Algorithms **3** (1982), 101–127.

39. O. Schirokauer, *On pro-finite groups and on discrete logarithms*, Ph. D. thesis, University of California, Berkeley, 68 pages, May 1992.

40. I. N. Stewart, D. O. Tall, *Algebraic number theory*, second edition, Chapman and Hall, London, 1987.

41. B. Vallée, *Generation of elements with small modular squares and provably fast integer factoring algorithms*, Math. Comp. **56** (1991), 823–849.

42. D. Wiedemann, *Solving sparse linear equations over finite fields*, IEEE Trans. Inform. Theory **32** (1986), 54–62.

43. H. G. Zimmer, *Computational problems, methods and results in algebraic number theory*, Lecture Notes in Math. **262**, Springer-Verlag, Berlin, 1972.

ROOM MRE-2Q334, BELLCORE, 445 SOUTH STREET, MORRISTOWN, NJ 07960, U.S.A.
E-mail address: lenstra@bellcore.com

DEPARTMENT OF MATHEMATICS, UNIVERSITY OF CALIFORNIA, BERKELEY, CA 94720, U.S.A.
E-mail address: hwl@math.berkeley.edu

DEC SRC, 130 LYTTON AVENUE, PALO ALTO, CA 94301, U.S.A.
E-mail address: msm@src.dec.com

TIDMARSH COTTAGE, MANOR FARM LANE, TIDMARSH, READING, BERKSHIRE, RG8 8EX, ENGLAND

THE LATTICE SIEVE

J. M. POLLARD

SUMMARY. We describe a possible improvement to the Number Field Sieve. In theory we can reduce the time for the sieve stage by a factor comparable with $\log(B_1)$. In the real world, where much factoring takes place, the advantage will be less. We used the method to repeat the factorisation of F_7 on an 8-bit computer (yet again!).

OBJECT OF THE LATTICE SIEVE

We will consider the case of F_7 throughout. We have

$$2F_7 = x^3 + 2, \qquad \text{where } x = 2^{43}.$$

The object of the sieve stage of the NFS is to find pairs of small coprime integers a and b such that:

i.   the integer $a + bx$ is smooth.
ii.   the polynomial $N(a, b) = a^3 - 2b^3$ is smooth.

This polynomial arises as the 'norm' of the algebraic integer $a + bx$, where x is a root of the equation $x^3 + 2 = 0$. For the sieve stage we don't need to know anything about algebraic numbers (for the whole algorithm we seem to need to know a little—more for general numbers).

At present everyone assumes [1, 2, 3] that the best thing to do is fix b (positive) and sieve over a range of values of a (positive and negative). We propose a different method, suggested by the 'special q' version of the QS method (due to Davis and Holdridge).

We divide the factor base into two parts:

S: the small primes: $p \leq B_0$,
M: the medium primes: $B_0 < p \leq B_1$.

We put $B_0/B_1 = k$, likely to be in the range 0.1–0.5; as usual, we also use:

L: the large primes: $B_1 < p \leq B_2$,

where B_2 is much larger than B_1.

Algorithm LS (the lattice sieve).

1.   Choose a region R of the (a, b) plane to be sieved.
2.   Choose a fixed prime q in M, the 'special prime', and sieve only those (a, b) pairs in R with

$$a + bx \equiv 0 \pmod{q}. \tag{1}$$

1991 *Mathematics Subject Classification.* Primary 11Y05, 11Y40.
Key words and phrases. Factoring algorithm, algebraic number fields.

The sieve is a double one, as usual:

i.   We sieve the numbers $a + bx$ with the primes $p < q$ only,

ii.   We sieve the numbers $N(a, b)$ with all the primes of S and M.

In both sieves we allow a large prime up to B_2. We discuss in a moment how the sieve works, but first we explain why it is (possibly) of interest.

Claim 1. The total number of integers sieved is much less than in the NFS.

In fact, it is reduced by the factor:

$$W = \sum_{q \in M} \frac{1}{q} \approx \frac{\log(1/k)}{\log(B_1)}. \tag{2}$$

Claim 2. We still get most of the solutions given by the NFS.

The ones we miss are those for which $a + bx$ has no prime factor in M. It is easiest to consider the case when no large prime is allowed in $(a + bx)$, that is, the *ff* and *fp* solutions of [1, 2]; then the fraction of solutions lost, L say, can be expressed in terms of Dickman's ρ function [4]. A random integer of size about bx has all factors $\leq B_1$ with probability:

$$\rho(\ln(bx)/\ln(B_1)), \tag{3}$$

and all factors $\leq B_0$ with probability:

$$\rho(\ln(bx)/\ln(B_0)). \tag{4}$$

The required fraction L is (4)/(3).

Example. The factorisation of F_9 [2].

Here we have $x = 2^{103}$, $b = 1.25(6)$ [middle of the range], $B_1 = 1.3(6)$. The table below compares:

$$W = \text{work done (eqn. 2), and}$$

$$L = \text{fraction of solutions lost,}$$

as functions of k.

k	W	$\ln(bx)/\ln(B_0)$	(4)	$L = (4)/(3)$
1.0	0.0	6.0687	1.71(−5)	1.0
0.5	0.0492	6.3830	6.38(−6)	0.373
0.4	0.0651	6.4912	4.52(−6)	0.264
0.3	0.0855	6.6363	2.85(−6)	0.167
0.2	0.1143	6.8521	1.42(−6)	0.083
0.1	0.1636	7.2554	3.81(−7)	0.022

Notes.

1. We used an approximate formula for ρ given by Pomerance [4] (ln is the natural logarithm):

$$\rho(u) = \exp(-u(\ln(u) + 0.56 - 1/\ln(u))), \qquad (5 < u < 11).$$

2. It may be that a larger proportion of the *pf* and *pp* solutions are lost.

Conclusion. An Infinitely Skilful Programmer (ISP) can get 83% of the solutions for 8.6% of the work.

OPERATION OF THE SIEVE

We assume that no primes of M divide x (true in our case, and in most cases of interest). The points (a, b) on (1) form a lattice $L(q)$ in the (a, b) plane. We begin by finding two short vectors which generate the lattice:

$$V_1 = (a_1, b_1) \quad \text{and} \quad V_2 = (a_2, b_2).$$

This is easy (my method resembles the extended gcd algorithm, applied to q and $-x$).

Then a typical point of the lattice is: $c \cdot V_1 + d \cdot V_2$, i.e.

$$(a, b) = (c \cdot a_1 + d \cdot a_2, \; c \cdot b_1 + d \cdot b_2). \tag{5}$$

We regard this as a point (c, d) in 'the (c, d) plane.'

For a and b to be coprime it is necessary that c and d be coprime. Is it sufficient? Not quite. It can happen that c and d are coprime and $a \equiv b \equiv 0 \pmod{q}$. In that case, a/q and b/q satisfy our conditions. So it suffices to consider coprime values of c and d.

We use a two-dimensional array $A[-C \ldots C, 1 \ldots D]$ whose $[c, d]$ element represents the point (5) (there is no point in allowing negative d, and the row $d = 0$ has only one useful element $A[1, 0]$ so we omitted that also). Note that (5) can produce negative values of b—then we change the signs of a and b ($b = 0$ is excluded, since then $a = \pm 1$, and (1) does not hold).

It seems sensible to choose $C > D$ for two reasons:

i.   because V_1 is shorter than V_2, sometimes much shorter.
ii.   because sieving is then more efficient (see below).

Clearly the use of a fixed rectangle in the (c, d) plane is crude, since this gives regions of different shapes in the (a, b) plane; but it seemed good enough for initial experiments.

We describe the working of the sieve in outline first, then in more detail. The array is used twice over, in the two stages of the sieve. In the first stage it is initially set to zero and then accumulates the sum of $\log(p)$ over those primes p which divide $(a + bx)$. We can find (easily) which values of $(a + bx)$ were smooth and these elements are marked provided that c and d are coprime.

The second stage is similar. The array is again set to zero and now accumulates sums of $\log(p)$ over those primes p which divide $N(a, b)$. Finally we find which elements $[c, d]$ have twice held smooth numbers and compute the values of a and b. From this point the algorithm is identical with the conventional NFS.

It remains to discuss how we sieve with a given number p, which is a prime or prime power. The two stages of the sieve are almost identical, so we consider the first stage. The array element $A[c, d]$ represents the integer:

$$a + bx = c \cdot u_1 + d \cdot u_2, \quad \text{where}$$
$$u_1 = a_1 + b_1 x \quad \text{and} \quad u_2 = a_2 + b_2 x.$$

Here of course $u_1 \equiv u_2 \equiv 0 \pmod{q}$, but u_1 and u_2 have no other prime factors in common, i.e. $(u_1, u_2) = q$.

Thus the element is to be sieved by p iff:

$$c \cdot u_1 + d \cdot u_2 \equiv 0 \pmod{p}. \tag{6}$$

Special cases arise when $u_1 \equiv 0 \pmod{p}$, or $u_2 \equiv 0 \pmod{p}$, when we must sieve the whole of every pth row, or every pth column.

In the general case, we either have $(u_1, p) = 1$ or $(u_2, p) = 1$ (both, unless p is a prime power); assume the first condition. There are at least two ways to proceed:

i. *Sieving by rows.* In each row each pth element is to be sieved. A convenient method is to sieve in both directions, starting from the least nonnegative solution of (6). We first calculate the inverse of $u_1 \pmod{p}$; then the starting value for each row is easily found from the last. This simple method is satisfactory for the smallest primes, but bad for the larger ones since no integers are to be sieved in many rows.

ii. *Sieving by vectors.* We make use of the fact that the points to be sieved form a lattice in the (c, d) plane (the corresponding points form a sublattice $L(pq)$ of $L(q)$ in the (a, b) plane). We compute two short vectors which generate the lattice:

$$v_1 = (c_1, d_1) \quad \text{and} \quad v_2 = (c_2, d_2).$$

Again, this is easy, but it must be done fast.

A typical point of the lattice is: $e \cdot v_1 + f \cdot v_2$, i.e.

$$(c, d) = (e \cdot c_1 + f \cdot c_2, \; e \cdot d_1 + f \cdot d_2).$$

We call this a point (e, f) in 'the (e, f) plane.' Again, it suffices to consider points with $(e, f) = 1$.

Problem. To generate quickly those points (e, f), preferably with e and f coprime, which give (c, d) points in the chosen rectangle.

In my version, described below, it suffices to consider small values of e and f. In a large scale implementation, sieving by vectors is essential, and must be done better.

More factorisations of F_7!

To compare the LS and conventional NFS methods we used the number $F_7 = 2^{128} + 1$, first factored by Brillhart and Morrison in 1970. Both algorithms were implemented on a Philips 2012, an 8-bit computer with 64k of memory and two 640k disc drives. A form of NFS was already run on this computer in 1988 [3], as a set of seven Pascal programs. The programs were subsequently improved to the method of [1]. Only one new program was needed for the LS (certainly capable of improvement).

The following applies to both methods. We have:

$$2F_7 = x^3 + 2, \quad \text{where } x = 2^{43}.$$

The first part of the factor base, FB_1, consists of the first 500 primes:

$$2, 3, 5, \ldots, 3571.$$

The second part FB_2 consists of algebraic primes of norm $\pm p$ arising in the factorisations of the rational primes of FB_1. There are 497 such primes. We included in FB_2 three units: -1, U and $1/U$ making 500 members in all. We have 81 'free' equations connecting FB_1 and FB_2 (and one more involving the units).

Our disc file of the factor base includes:

i.   For each prime p of FB_1, the residue x (mod p).
ii.   For each member of FB_2, of norm p, the corresponding root r of:

$$r^3 \equiv -2 \pmod{p}. \tag{7}$$

Indeed, for our present purposes, FB_2 is a list of the roots of (7) for p in FB_1.

Our versions of both methods allowed a large prime in $a + bx$, but not in $N(a, b)$; thus we used only ff and pf solutions.

1. The conventional Number Field Sieve.
The choice of parameters given here is better than in [3].

$$
\begin{aligned}
\text{Range of } b = \quad & 1 \ldots 1350 \\
\text{Range of } a = \ & -6000 \ldots 6000 \\
\text{Limit for large prime} = \quad & 30{,}000 \\
\text{Integers sieved} = \quad & 1.62(7) \\
ff \text{ solutions} = \quad & 480 \\
pf \text{ solutions} = \quad & 1941 \\
pf \text{ pairs} = \quad & 484 \\
\text{Free solutions} = \quad & 81 \\
\text{Total solutions} = \quad & 1045 \\
\text{Integers sieved per } ff \text{ solution} = \ & 33{,}750 \\
\text{Integers sieved per } pf \text{ solution} = \ & 8{,}350
\end{aligned}
$$

2. The Lattice Sieve (sieving by rows for all p).

The first 142 primes were small   : $2, 3, 5, \ldots, 821$.
The next 358 primes were medium: $823, \ldots, 3571$.
(Thus $k = 821/3571 = 0.230$).

$$
\begin{aligned}
\text{Limit for large prime} = \quad & 30{,}000 \\
\text{Dimensions of sieve array:} \quad & [-100 \ldots 100, 1 \ldots 60] \\
\text{Integers sieved per } q = \quad & 12{,}060 \\
\text{Total integers sieved} = \quad & 4.32(6) \\
ff \text{ solutions} = \quad & 510 \\
pf \text{ solutions} = \quad & 1732 \\
pf \text{ pairs} = \quad & 418 \\
\text{Free solutions} = \quad & 81 \\
\text{Total solutions} = \quad & 1009 \\
\text{Integers sieved per } ff \text{ solution} = \ & 8{,}466 \\
\text{Integers sieved per } pf \text{ solution} = \ & 2{,}493
\end{aligned}
$$

Both methods succeeded in factoring F_7. We see that even this crude version of
LS reduces the number of integers sieved by a factor of 4. At present the LS is
30% slower (6.5 hours compared to 4.9 hours). Some problems are:

i.   Multiplication and division are slow on 8-bit computers; we need them for
       gcd, inverse (mod p) and shortest vector calculations (these are all rather
       similar, and can perhaps be done by the 'binary gcd' method).

ii.   Accessing two-dimensional arrays is also slow (perhaps we should only use
       one-dimensional arrays).

Some programming notes

1. My main array A is of 8-bit elements. I use a second array B to denote the
status of the elements of the A. The elements of B take only three values, so I
need only 2 bits (but currently use 8). Initially we have:

$$B[c, d] = \begin{cases} 0, & \text{if } c \text{ and } d \text{ not coprime,} \\ 1, & \text{if } c \text{ and } d \text{ are coprime.} \end{cases}$$

After the first sieve, only the 1-elements are examined, and some of the '1's are
changed to '2's:

$$B[c, d] = 2, \quad (c, d) \text{ passed through first sieve.}$$

After the second sieve, only the 2-elements are examined, and the '2's changed
back to '1's.

2. After the first sieve, we must compare

$$A[c, d] \quad \text{and} \quad \text{'log}(a + bx) + \text{constant'}$$

where b is variable, unlike the NFS. To save many logarithms, we take exponen-
tials and compare instead:

$$\text{Table}[A[c, d]] \quad \text{and} \quad \text{abs}(b).$$

3. Optionally, we can add $\log(p)$ to $A[c, d]$ only when that element is still interesting, i.e. marked '1' in the first sieve, or '2' in the second. In Pascal, this is
worthwhile for the second sieve at least (we access $A[c, d]$ twice in adding to it).

4. In my program, sieving by vectors gains very little time, but I recently made
an attempt in this direction, since it is an essential part of the algorithm. My
present method is to allow only $e, f = -2, \ldots, 2$, so that there are just nine
(e, f) points to try:

$$(1, 0), (0, 1), (1, 1), (2, 1), (1, 2), (-1, 1), (-1, 2), \pm(-2, 1). \tag{8}$$

This means that some (c, d) points get missed (see below).

Numerical Example from the Factorisation of F_7

Take special prime $q = 3571$ (the 500th and largest prime of FB_1). We find

$$V_1 = (-10, 29) \quad \text{and} \quad V_2 = (99, 70).$$

An example of an *ff* solution is:

$$(c, d) = (-43, 6),$$
$$(a, b) = (-1024, 827) \quad \text{(signs changed to make } b \text{ positive)},$$
$$a + bx = 2^{10} \cdot 3^2 \cdot 7 \cdot 13 \cdot 1301 \cdot 1867 \cdot 3571,$$
$$a^3 - 2b^3 = -2 \cdot 5 \cdot 59 \cdot 1789 \cdot 2089.$$

To sieve by vectors with $p = 1867$, we begin by calculating:

$$v_1 = (-43, 6) \quad \text{and} \quad v_2 = (46, 37).$$

We find that 5 of the 9 points in (8) give (c, d) points in our rectangle:

(e, f)	(c, d)	
$(1, 0)$	$(-43, 6)$	(gives *ff* solution above)
$(0, 1)$	$(46, 37)$	
$(1, 1)$	$(3, 43)$	
$(-1, 1)$	$(89, 31)$	
$(2, 1)$	$(-40, 49)$	

Sieving by rows reveals that we missed one point:

$$(3, 1) \qquad (-83, 55)$$

If we sieve by vectors for the upper 250 primes of M only, (as in Note 4), we miss 7 out of 517 points in the first sieve, and 2 out of 28 in the second.

References

1. A. K. Lenstra, H. W. Lenstra, Jr., M. S. Manasse, J. M. Pollard, *The number field sieve*, this volume, pp. 11–42; extended abstract: Proc. 22nd Annual ACM Symp. on Theory of Computing (STOC), Baltimore, May 14–16, 1990, 564–572.
2. A. K. Lenstra, H. W. Lenstra, Jr., M. S. Manasse, J. M. Pollard, *The factorization of the ninth Fermat number*, Math. Comp. 61 (1993), to appear.
3. J. M. Pollard, *Factoring with cubic integers*, unpublished manuscript, 1988; this volume, pp. 4–10.
4. C. Pomerance, *Factoring*, pp. 27–47 in: C. Pomerance (ed.), *Cryptology and computational number theory*, Proc. Sympos. Appl. Math. **42**, Amer. Math. Soc., Providence, 1990.

Tidmarsh Cottage, Manor Farm Lane, Tidmarsh, Reading, Berkshire, RG8 8EX, England

FACTORING INTEGERS WITH THE NUMBER FIELD SIEVE

J. P. BUHLER, H. W. LENSTRA, JR., CARL POMERANCE

ABSTRACT. In 1990, the ninth Fermat number was factored into primes by means of a new algorithm, the "number field sieve", which was proposed by John Pollard. The present paper is devoted to the description and analysis of a more general version of the number field sieve. It should be possible to use this algorithm to factor arbitrary integers into prime factors, not just integers of a special form like the ninth Fermat number. Under reasonable heuristic assumptions, the analysis predicts that the time needed by the general number field sieve to factor n is $\exp((c+o(1))(\log n)^{1/3}(\log \log n)^{2/3})$ (for $n \to \infty$), where $c = (64/9)^{1/3} \doteq 1.9223$. This is asymptotically faster than all other known factoring algorithms, such as the quadratic sieve and the elliptic curve method.

1. INTRODUCTION

In 1988 John Pollard circulated a manuscript [31] that described a new method for factoring integers. The procedure required the use of an algebraic number field tailored for the specific number n to be factored. In [24] a practical version of this idea was presented, dubbed by the authors "the number field sieve". This method has had several noteworthy successes in factoring numbers of the form $n = b^c \pm 1$, where b is small, from the Cunningham project (see [5]). The most spectacular of these factorizations was that of the ninth Fermat number $F_9 = 2^{2^9} + 1$, which has 155 decimal digits (see [23]).

The number field sieve has, so far, only been applied to factor numbers where certain desiderata were easily met. These include a monic irreducible polynomial $f \in \mathbf{Z}[X]$ of "small, but not too small" degree d, with "small" coefficients, and an integer $m \approx n^{1/d}$ such that $f(m) \equiv 0 \bmod n$. Further, if α is a zero of f, it is convenient for the ring of integers $\mathcal{O}$ of the number field $K = \mathbf{Q}(\alpha)$ to be not too much larger than $\mathbf{Z}[\alpha]$, for $\mathcal{O}$ to have class number one, and for the units of $\mathcal{O}$ to be easily computable.

For example, in the case $n = F_9$ the polynomial $f = X^5 + 8$ and the integer $m = 2^{103}$ were used; note that $f(m) = m^5 + 8 = 2^{515} + 8 \equiv 0 \bmod n$. More generally, for several numbers $n = b^c \pm 1$, with b small and c large, it has been fairly easy to meet the list of desiderata and to use the number field sieve to

1991 *Mathematics Subject Classification.* Primary 11Y05, 11Y40.

Key words and phrases. Factoring algorithm, algebraic number fields.

The authors wish to thank Dan Bernstein, Arjeh Cohen, Michael Filaseta, Andrew Granville, Arjen Lenstra, Victor Miller, Robert Rumely, and Robert Silverman for their helpful suggestions. The authors were supported by NSF under Grants No. DMS 90-12989, No. DMS 90-02939, and No. DMS 90-02538, respectively. The second and third authors are grateful to the Institute for Advanced Study (Princeton), where part of the work on which this paper is based was done.

factor n. For numbers of this form it was suggested in [24] that the number field sieve takes time at most $L_n[\frac{1}{3}, (32/9)^{1/3} + o(1)]$ to factor n as n goes to infinity, where

$$L_n[u, v] = \exp(v(\log n)^u (\log \log n)^{1-u}).$$

The exponent $u = \frac{1}{3}$ in the number field sieve is the new and exciting aspect of this complexity function since all other known algorithms, such as the quadratic sieve or the elliptic curve method, have complexity, heuristic or probabilistic, at least $L_n[\frac{1}{2}, 1 + o(1)]$ for n tending to infinity through an infinite sequence of numbers.

Can the number field sieve be extended to general integers? It is to this question that this paper is addressed. We show that the method can be modified so that an arbitrary integer n can be factored with heuristic complexity $L_n[\frac{1}{3}, (64/9)^{1/3} + o(1)]$ for $n \to \infty$. We will call the new algorithm the number field sieve; if we need to specifically refer to the earlier algorithm we will refer to it as the special number field sieve.

The reason the constant $(64/9)^{1/3} \doteq 1.922999$ for the general case is larger than the constant $(32/9)^{1/3} \doteq 1.526285$ for the special number field sieve is that the coefficients of the polynomial f we construct below are about $n^{1/d}$. This is in a rough sense asymptotically best possible for general n, as we shall see in 12.10. For special values of n it may be possible to choose the coefficients of f much smaller, which makes the algorithm faster.

Is the number field sieve practical? Since it involves the same underlying sieving operations as, for instance, the quadratic sieve and the special number field sieve, it is our guess that this algorithm will eventually be the method of choice for sufficiently large integers. At the moment, its crossover with the "state-of-the-art" algorithm for factoring, namely the quadratic sieve, seems to be about 125 digits. This is so high that it is very difficult to factor a general number of this size with either method. The current record with the quadratic sieve is 116 decimal digits (see [25]). However, time is on the side of the number field sieve. It is reasonable to expect that hardware will improve and that the number field sieve will be refined and polished as it becomes better understood. Of course it is impossible to predict the future; some other faster factoring algorithm may be discovered that will supplant the quadratic sieve before theoretical and practical advances give the number field sieve its day in the sun.

If we compare the relative predicted performance of the number field sieve and the quadratic sieve on the basis of the somewhat questionable assumption that the "$o(1)$" terms in the heuristic complexity estimates can be ignored, then we find that the predicted number of operations for both are within a factor of about 3 for numbers between 100 and 150 decimal digits. This suggests that a small change in the implementation of either algorithm may have a large effect on the location of the crossover point.

Our description of the number field sieve incorporates the idea of Adleman [1] of using 'character columns', described in Section 8. In our original formulation of the number field sieve we had used a more awkward technique instead of character columns, which initially achieved only $L_n[\frac{1}{3}, 9^{1/3} + o(1)]$ as $n \to \infty$ for the heuristic complexity of the number field sieve, where $9^{1/3} \doteq 2.080084$; and it was only at the expense of considerable additional complications that we could obtain the bound $L_n[\frac{1}{3}, (64/9)^{1/3} + o(1)]$ with this technique. Adleman's idea achieves the latter bound with much less effort, and it simplifies the description of the algorithm in several ways. In addition it likely moves the number field sieve closer to being a practical factoring algorithm for arbitrary integers.

Another improvement to be mentioned is that of Coppersmith [10]. His idea reduces the complexity estimate even further, namely to $L_n[\frac{1}{3}, c + o(1)]$ for $n \to \infty$, where

$$c = \frac{(92 + 26\sqrt{13})^{1/3}}{3} \doteq 1.901884.$$

However, it is unlikely that this method will be practical for numbers of reasonable size (of fewer than 1000 digits, say).

The idea underlying the number field sieve has also been applied to the discrete logarithm problem. For this, we refer to [15] and [35].

The structure of this paper is as follows. Section 2 contains an outline of the number field sieve. In Section 3 we describe an algorithm for selecting the number field to be used by the algorithm. Section 4 is devoted to a description of a well-known sieving technique for constructing squares in the field of rational numbers. In Section 5 we carry this technique over to the algebraic number field. It turns out that we have to deal with certain obstructions, which are described and analyzed in Section 6. Two algebraic facts that are used in Sections 5 and 6 are proved in Section 7. We overcome the obstructions in Section 8, by using the character columns that were suggested by Adleman. In Section 9 we discuss a problem that has not appeared in earlier factoring algorithms, namely that of taking square roots in algebraic number fields. In Section 10 we state a heuristic principle that can be used to obtain running time estimates for a surprisingly wide class of factoring algorithms. Section 11 summarizes the entire algorithm and gives a heuristic running time analysis. Finally, in Section 12 we describe a modification of the number field sieve that should improve its practical performance.

2. THE IDEA OF THE NUMBER FIELD SIEVE

A very old factoring strategy going back to Fermat and Legendre is to write n as a difference of two squares. More generally, it suffices to find a solution to $x^2 \equiv y^2 \mod n$. One might then obtain a factorization of n by finding the greatest common divisor of $x - y$ and n. In fact, it is easy to prove that if n is divisible by at least two distinct odd primes then for at least half of the pairs $x \mod n$, $y \mod n$ with $x^2 \equiv y^2 \mod n$ and $\gcd(xy, n) = 1$, we have $1 < \gcd(x - y, n) < n$. There are many factoring algorithms that exploit this idea by trying to construct such pairs x, y in a random or pseudo-random manner. These algorithms include the continued fraction method [30], the random squares method [12], the quadratic sieve [33], and, of course, the special number field sieve.

Before we see how the number field sieve attempts to find a solution to $x^2 \equiv y^2 \mod n$ we say a few words about the ring in which the number field sieve operates. Suppose $f \in \mathbf{Z}[X]$ is monic and irreducible of degree $d > 1$. We shall work with the ring $\mathbf{Z}[\alpha]$ that is generated by a zero α of f. It makes no difference whether one thinks of $\mathbf{Z}[\alpha]$ as a subring of the field of complex numbers or as the ring $\mathbf{Z}[X]/f\mathbf{Z}[X]$, with $\alpha = (X \bmod f)$; all that matters is that each element of $\mathbf{Z}[\alpha]$ can in a unique way be written in the form $\sum_{i=0}^{d-1} a_i \alpha^i$, with $a_0, a_1, \ldots, a_{d-1} \in \mathbf{Z}$. Thus, each element of $\mathbf{Z}[\alpha]$ can be represented as a vector with d integral coordinates a_i. The addition in the ring is then just vector addition. To multiply two polynomial expressions in α, one first multiplies them as polynomials, and next uses the relation $f(\alpha) = 0$ to reduce the result to a polynomial expression of degree less than d in α. If we let, in a completely analogous way, the a_i range over the field $\mathbf{Q}$ of rational numbers rather than over $\mathbf{Z}$, then we obtain the field of fractions $\mathbf{Q}(\alpha)$ of $\mathbf{Z}[\alpha]$.

Coming back to the number field sieve, let us now assume that $m \in \mathbf{Z}$ satisfies $f(m) \equiv 0 \bmod n$. Then there is a natural ring homomorphism $\varphi \colon \mathbf{Z}[\alpha] \to \mathbf{Z}/n\mathbf{Z}$ induced by $\varphi(\alpha) = (m \bmod n)$; so $\varphi(\sum_i a_i \alpha^i) = (\sum_i a_i m^i \bmod n)$. Suppose we can find a non-empty set S of pairs (a, b) of relatively prime integers with the following two properties:

$$(2.1) \qquad \qquad \prod_{(a,b) \in S} (a + bm) \quad \text{is a square in } \mathbf{Z},$$

$$(2.2) \qquad \qquad \prod_{(a,b) \in S} (a + b\alpha) \quad \text{is a square in } \mathbf{Z}[\alpha].$$

Let $x \in \mathbf{Z}$ be a square root of the square in (2.1) and let $\beta \in \mathbf{Z}[\alpha]$ be a square root of the element of $\mathbf{Z}[\alpha]$ in (2.2). Since $\varphi(a + b\alpha) = (a + bm \bmod n)$, we have $\varphi(\beta^2) = (x^2 \bmod n)$. Let $y \in \mathbf{Z}$ be such that $\varphi(\beta) = (y \bmod n)$. Then $y^2 \equiv x^2 \bmod n$, and we have constructed our congruent squares and so may attempt to factor n by computing $\gcd(x - y, n)$.

There are several questions that are raised by the above outline:

(i)   How are the polynomial f and the integer m to be constructed?

(ii) How is the set S of coprime integer pairs that satisfies (2.1) and (2.2) to be found?

(iii) How is an element $\beta \in \mathbf{Z}[\alpha]$ to be found such that β^2 is the square in (2.2)?

(iv) How much time do these steps take?

The overall plan of this paper is to gradually answer these questions until we can finally state a precise version of the algorithm and attempt to analyze its complexity.

Remark. The basic goal of most "combination of congruences" factoring algorithms, including the number field sieve, can be encapsulated algebraically in the following way. We have a ring R and a ring homomorphism $\psi \colon R \to \mathbf{Z}/n\mathbf{Z} \times \mathbf{Z}/n\mathbf{Z}$ together with a means of generating many elements of R whose image under ψ lies in the diagonal $\{(x, x) : x \in (\mathbf{Z}/n\mathbf{Z})^*\}$. We then hope to combine these elements multiplicatively to obtain squares in R whose square roots have an image under ψ not lying in $\{(x, \pm x) : x \in (\mathbf{Z}/n\mathbf{Z})^*\}$. In the case of the quadratic sieve we have $R = \mathbf{Z} \times \mathbf{Z}$. In the case of the number field sieve we have $R = \mathbf{Z} \times \mathbf{Z}[\alpha]$ and $\psi(r, \beta) = (r \bmod n, \varphi(\beta))$, and we consider elements of the form $(a+bm, a+b\alpha)$. It is tempting to consider more general rings, e.g., $R = \mathbf{Z}[\alpha] \times \mathbf{Z}[\alpha']$, or $R = \mathbf{Z}[\alpha]$ where f has *two* zeroes modulo n, but so far we have not found a way to exploit this.

3. Finding a Polynomial

Given a positive integer n that is not a prime power, the first step of the number field sieve algorithm is to find a polynomial f with integer coefficients and an integer m such that $f(m)$ is a multiple of n. In the basic version of the number field sieve that we will present, the following particularly simple method is used to find a polynomial; this algorithm will be referred to as the "base m" method.

Suppose that we are given positive integers n and d with $d > 1$ and $n > 2^{d^2}$. Set $m = [n^{1/d}]$, and write n to the base m:

$$(3.1) \qquad n = c_d m^d + c_{d-1} m^{d-1} + \cdots + c_0$$

where the "digits" c_i satisfy, as usual, the inequality $0 \leq c_i < m$. The output of the base m algorithm consists of the integer m and the polynomial $f = c_d X^d + c_{d-1} X^{d-1} + \cdots + c_1 X + c_0$. Note that we have $f(m) = n$.

Proposition 3.2. *The leading coefficient c_d of f is equal to 1, and $c_{d-1} \leq d$.*

Proof. From our assumption $n > 2^{d^2}$ we have $\binom{d}{i} \leq 2^d - 2 \leq n^{1/d} - 2 \leq m - 1$. Therefore the digits of $(m + 1)^d$ in the base m are the binomial coefficients $\binom{d}{i}$, and the proposition follows from the inequalities $m^d \leq n < (m + 1)^d$.

For the d that we will recommend later, n will be much larger than 2^{d^2}.

The polynomial f produced by the base m algorithm may be reducible. However, since our interest lies in factoring n, this event would be fortunate. Indeed, if $f = gh$ is a non-trivial factorization of f in $\mathbf{Z}[X]$ then $g(m)h(m) = f(m) = n$ is a non-trivial splitting of n in $\mathbf{Z}$. This result follows from the proofs in [4], where we need only the easier case $m \geq 3$. We note that f can be factored in time $(\log n)^{O(1)}$, by means of the algorithm of [22].

In a weak asymptotic sense, the base m algorithm, simple as it may be, cannot be improved for use in the number field sieve, although for practical purposes there is still room for improvement. This is further discussed in 12.10 and 12.15.

The following estimate will be needed later in this paper. We let f be as produced by the base m algorithm, with $d > 1$, $n > 2^{d^2}$.

Lemma 3.3. *The discriminant Δ of f satisfies $|\Delta| < d^{2d} n^{2-3/d}$.*

Proof. The discriminant of the monic polynomial f is, up to sign, equal to the resultant of f and its derivative, which in turn can be expressed as the determinant of the corresponding Sylvester matrix (see [37, Sections 34 and 35]). The non-zero entries of each of the first $d - 1$ rows of that matrix are the coefficients of f, and the non-zero entries of each of the remaining d rows are the coefficients of f'. To estimate the determinant, we divide each of the last d rows, corresponding to f', by d, and we divide each of the last $2d - 3$ columns by m; those are the columns involving a c_i with $i < d - 1$. Finally, we subtract c_{d-1} times the first column from the second column. This results in a matrix of which all entries are at most 1 in absolute value. Each of the first $d - 1$ row vectors of that matrix has Euclidean length at most $\sqrt{d+1}$, and each of the last d row vectors has Euclidean length at most $\sqrt{d}$. Thus from Hadamard's determinant bound we obtain

$$|\Delta| \leq d^d m^{2d-3}(d+1)^{(d-1)/2} d^{d/2} < d^{2d} n^{2-3/d},$$

using $m^d \leq n$ and $d > 1$ for the last inequality. This proves 3.3.

4. THE RATIONAL SIEVE

We let n and d be integers with n, $d > 1$, and we let $f \in \mathbf{Z}[X]$ be a monic irreducible polynomial of degree d. We let m be an integer with the property $f(m) \equiv 0 \bmod n$. By α we denote a zero of f, as explained in Section 2. We write $\mathbf{Z}[\alpha]$ for the ring generated by α.

As we saw above, the heart of the number field sieve lies in constructing a non-empty set S of coprime integer pairs for which we have

(4.1) $$\prod_{(a,b) \in S} (a + bm) \quad \text{is a square in } \mathbf{Z},$$

(4.2) $$\prod_{(a,b) \in S} (a + b\alpha) \quad \text{is a square in } \mathbf{Z}[\alpha].$$

Basically, the construction of S proceeds in two steps. First, one uses a sieve to find a set T of pairs (a, b) such that both $a + bm$ is smooth (i.e., factors into small primes), and $a + b\alpha$ is smooth (in a similar sense, to be defined later) in $\mathbf{Z}[\alpha]$. Next, one uses linear algebra over the field with two elements to locate $S \subset T$.

Let u be a large positive number to be chosen later, depending on n. Our overall universe of possible pairs, from which the sets T and S will be chosen, is

(4.3) $$U = \{(a, b) : a, b \in \mathbf{Z},\ \gcd(a, b) = 1,\ |a| \leq u,\ 0 < b \leq u\}.$$

We will need to choose the parameter u sufficiently large so that U contains a non-empty set S satisfying (4.1) and (4.2).

Initially, we will discuss conditions (4.1) and (4.2) separately. That is, in the present section we focus on the "rational" side of the number field sieve, i.e., finding a set S satisfying (4.1). Next we shall concentrate on the "algebraic" side (4.2). Finally, we shall see how to achieve (4.1) and (4.2) simultaneously.

The procedure for finding a square in $\mathbf{Z}$ by sieving is standard; we recall the idea. First a parameter $y = y(n)$ is chosen, and by sieving one finds a subset

$$T_1 = \{(a, b) \in U : a + bm \text{ is } y\text{-smooth}\},$$

where we say that an integer is y-smooth if all of its prime divisors are less than or equal to y. The sieving procedure works as follows. For each fixed integer b with $0 < b \leq u$ an array is initialized with the integers $a + bm$ for $-u \leq a \leq u$. For each prime number $p \leq y$ the numbers in the array corresponding to values of a with $a \equiv -bm \bmod p$ are retrieved one at a time, divided by the highest power of p that divides them, and the quotient is replaced in the same array at the same location from which the number was retrieved. At the end of this procedure the number in the ath location is, up to sign, the largest divisor of $a + bm$ that is coprime to the primes up to y. Any location that contains the number 1 or -1 at the end of the procedure corresponds to a number $a + bm$ that is y-smooth. If $\gcd(a, b) = 1$, we have thus detected a member of T_1.

In practice various devices can be used to speed up the sieving. For instance, it is more efficient to replace the numbers in the array by their approximate logarithms (say to base 2), to initialize the array with 0 instead of the logarithms of the numbers $|a + bm|$, to add the logarithm of p instead of dividing by p, to ignore small primes, to ignore higher powers of p, and to inspect, at the end of the procedure, all values of a for which the ath location contains a number exceeding a certain bound independent of a.

Remark. The primes less than or equal to y are said to be in the "factor base" of the sieve. The precise choice of the parameters y and u will be given later as part of the complexity analysis of the final algorithm, see Section 11.

Suppose the parameters y and u are chosen so that $\#T_1 > \pi(y) + 1$, where $\#$ denotes cardinality and $\pi(y)$ denotes the number of primes up to y. It is well-known that by using linear algebra over the field $\mathbf{F}_2$ with two elements one can find a non-empty subset S of T_1 for which (4.1) holds; again we recall the idea.

Let $B = \pi(y)$, let p_j denote the jth prime, for $1 \leq j \leq B$, and let $p_0 = -1$. For a y-smooth integer

$$w = \prod_{j=0}^{B} p_j^{e_j}$$

we define the exponent vector $e(w) \in \mathbf{F}_2^{B+1}$ by

$$e(w) = (e_0 \bmod 2, e_1 \bmod 2, \ldots, e_B \bmod 2).$$

We may form such a vector $e(a + bm)$ for each $(a, b) \in T_1$. Since the number of such vectors exceeds the dimension of the $\mathbf{F}_2$-vector space $\mathbf{F}_2^{B+1}$, there is a non-trivial linear dependence relation with coefficients 0 and 1, and hence a non-empty subset $S \subset T_1$ such that

$$\sum_{(a,b) \in S} e(a + bm) = 0 \in \mathbf{F}_2^{B+1}.$$

Therefore

$$\prod_{(a,b) \in S} (a + bm) \quad \text{is a square in } \mathbf{Z}.$$

Thus we have "solved" (4.1) by combining smooth elements.

5. The algebraic sieve

The notation and hypotheses in this section are as in Section 4. In addition, we write K for the field of fractions $\mathbf{Q}(\alpha)$ of $\mathbf{Z}[\alpha]$ (see Section 2) and $\mathcal{O}$ for the ring of algebraic integers in K. The multiplicative group of K is indicated by K^*, and $N \colon K \to \mathbf{Q}$ is the norm map of the extension $\mathbf{Q} \subset K$. For background on algebraic number theory we refer to [19; 40].

In order to find a square in $\mathbf{Z}[\alpha]$, i. e., find a set satisfying (4.2), we attempt to mimic the well-worn strategy described in the previous section. If the ring $\mathbf{Z}[\alpha]$ is a unique factorization domain this would be fairly easy, though problems with units would still remain. We note that in only a few of the applications so far of the special number field sieve, $\mathbf{Z}[\alpha]$ has been a unique factorization domain, but in the remaining cases where it has not, the full ring of integers $\mathcal{O}$ in K has been. Since we certainly cannot count on this being true for arbitrary numbers, we will describe a strategy for solving (4.2) that does not depend on special properties of $\mathbf{Z}[\alpha]$.

Define an element $\beta \in \mathbf{Z}[\alpha]$ to be y-smooth if its norm $N(\beta) \in \mathbf{Z}$ is y-smooth. We can calculate the norm of an element of the form $a + b\alpha$ by substituting a, b for X, Y in the homogeneous polynomial $(-Y)^d f(-X/Y)$; that is, if $a, b \in \mathbf{Z}$ then

$$(5.1) \qquad N(a + b\alpha) = a^d - c_{d-1} a^{d-1} b + \cdots + (-1)^d c_0 b^d$$

where $f = X^d + c_{d-1} X^{d-1} + \cdots + c_0$.

A modification of the earlier sieving idea can be used to find the set

$$T_2 = \{(a, b) \in U : a + b\alpha \text{ is } y\text{-smooth}\},$$

where U is as in (4.3). Namely, for each prime p let the set of zeroes of f mod p be denoted by $R(p)$, i. e., $R(p) = \{r \in \{0, 1, \ldots, p-1\} : f(r) \equiv 0 \bmod p\}$. Then for any fixed integer b with $0 < b \le u$ and $b \not\equiv 0 \bmod p$, the integers a with $N(a + b\alpha) \equiv 0 \bmod p$ are those with $a \equiv -br \bmod p$ for some $r \in R(p)$. Note that if $b \equiv 0 \bmod p$, then there are no integers a with $(a, b) \in U$ and $N(a + b\alpha) \equiv 0 \bmod p$.

For each fixed b initialize an array with the numbers $N(a+b\alpha)$ for $-u \le a \le u$. For each prime $p \le y$ that does not divide b and each choice of $r \in R(p)$ the positions corresponding to a that are congruent to $-br \bmod p$ are identified, the numbers in these positions are retrieved and divided by the highest power of p that divides them and then the quotient is replaced in the array as before. At the end of this process the locations containing ± 1 correspond to y-smooth values of $a + b\alpha$ with $\gcd(a, b) = 1$, and hence to elements of T_2. We can make this procedure more efficient by using the techniques mentioned in the previous section, including the use of approximate logarithms.

Remark 5.2. Note that for each prime p we might sieve as many as d residue classes modulo p; however, heuristically the average size of $R(p)$ is about 1 (see [19, Chapter VIII, Section 4]). (This would even be provable if we were to choose y large enough.)

The next step is to apply linear algebra over the field with two elements, but here some complications arise. In the previous section we combined the numbers $a + bm$, for $(a, b) \in T_1$, into a square by using their exponent vectors. Similarly, we can now use the exponent vectors of the numbers $N(a + b\alpha)$ for $(a, b) \in T_2$ and proceed with them in the same way. However, this leads to a subset $S \subset T_2$ for which only the *norm* of the product $\prod_{(a,b) \in S} (a + b\alpha)$ is a square (in $\mathbf{Z}$). This is a necessary condition for the product *itself* to be a square in $\mathbf{Z}[\alpha]$ (or even just in K), but it is very far from being sufficient. It turns out that we can overcome this problem almost completely by keeping track, for each prime number p dividing $N(a + b\alpha)$, of the value $r \in R(p)$ that is "responsible" for the fact that p divides $N(a + b\alpha)$.

More explicitly, let a, $b \in \mathbf{Z}$ satisfy $\gcd(a, b) = 1$. Further let p be a prime number and r an element of the set $R(p)$ defined above. Then we define $e_{p,r}(a + b\alpha)$ by

$$e_{p,r}(a + b\alpha) = \begin{cases} \mathrm{ord}_p(N(a + b\alpha)) & \text{if } a + br \equiv 0 \bmod p \\ 0 & \text{otherwise,} \end{cases}$$

where $\mathrm{ord}_p(k)$ is the number of factors p in k. Clearly we have

$$N(a + b\alpha) = \pm \prod_{p,r} p^{e_{p,r}(a+b\alpha)},$$

the product ranging over all pairs p, r with p prime and $r \in R(p)$. The following result justifies the introduction of the numbers $e_{p,r}(a + b\alpha)$.

Proposition 5.3. *Let S be a finite set of coprime integer pairs (a, b) with the property that $\prod_{(a,b) \in S} (a + b\alpha)$ is the square of an element of K. Then for each prime number p and each $r \in R(p)$ we have*

$$\sum_{(a,b) \in S} e_{p,r}(a + b\alpha) \equiv 0 \bmod 2.$$

This proposition is proved below.

For the number field sieve we are really interested in the converse of the proposition: if the congruence in 5.3 holds for all pairs p, r, does it follow that $\prod_{(a,b) \in S} (a + b\alpha)$ is a square? The answer is "no", as is shown by the example $S = \{(-1, 0)\}$, if K does not contain a square root of -1. However, we shall see, using the results in Section 7, that the extent to which the converse fails can be measured, that it is quite small (see Theorem 6.7), and that the failure of the converse can be overcome by the use of quadratic characters (see Section 8).

In order to prove 5.3 it is convenient to recall some basic facts about the non-zero prime ideals, or "primes" as we shall call them, of the ring $\mathbf{Z}[\alpha]$. If $\mathfrak{p} \subset \mathbf{Z}[\alpha]$ is a prime, then $\mathbf{Z}[\alpha]/\mathfrak{p}$ is a finite field, and $\mathfrak{p}$ contains a unique prime number p (see Section 7). The *norm* $\mathfrak{N}\mathfrak{p}$ of a prime $\mathfrak{p}$ is the number of elements $\mathfrak{N}\mathfrak{p} = \#\mathbf{Z}[\alpha]/\mathfrak{p}$ of its residue class field, and the *degree* of $\mathfrak{p}$ is the degree of $\mathbf{Z}[\alpha]/\mathfrak{p}$ as a field extension of its prime field $\mathbf{F}_p$. If $\mathfrak{p}$ is a *first degree* prime, then

$\mathbf{Z}[\alpha]/\mathfrak{p}$ is isomorphic to $\mathbf{F}_p$, we have $\mathfrak{N}\mathfrak{p} = p$, and the map $\mathbf{Z}[\alpha] \to \mathbf{F}_p$ with kernel $\mathfrak{p}$ sends α to a zero $r \bmod p$ of $f \bmod p$. Hence, a first degree prime $\mathfrak{p}$ gives rise to a pair p, r as considered above. Conversely, if p is a prime number and $r \in R(p)$, then there is a unique ring homomorphism $\mathbf{Z}[\alpha] \to \mathbf{F}_p$ that maps α to $r \bmod p$, and its kernel is a first degree prime $\mathfrak{p}$ of $\mathbf{Z}[\alpha]$. Thus there is a one-to-one correspondence between pairs p, r with $r \in R(p)$ and first degree primes $\mathfrak{p} \subset \mathbf{Z}[\alpha]$; the ideal $\mathfrak{p}$ corresponding to p, r is generated by p and $\alpha - r$.

We shall interpret the number $e_{p,r}(a + b\alpha)$ defined above as the "number of factors $\mathfrak{p}$ in $a + b\alpha$", where $\mathfrak{p}$ corresponds to p, r. If $\mathbf{Z}[\alpha]$ is equal to the full ring of integers $\mathcal{O}$ of K then it is clear what we mean by this: it is a standard fact from algebraic number theory that non-zero ideals of $\mathcal{O}$ factor uniquely into primes, and $e_{p,r}(a + b\alpha)$ is the exponent of $\mathfrak{p}$ in the factorization of the ideal $(a + b\alpha)\mathcal{O}$. In order to generalize this to the case in which $\mathbf{Z}[\alpha] \neq \mathcal{O}$ we need the following result.

Proposition 5.4. *There is, for each prime $\mathfrak{p}$ of $\mathbf{Z}[\alpha]$, a group homomorphism $l_{\mathfrak{p}} \colon K^* \to \mathbf{Z}$, such that the following hold:*
(a) $l_{\mathfrak{p}}(\beta) \geq 0$ for all $\beta \in \mathbf{Z}[\alpha]$, $\beta \neq 0$;
(b)   if $\beta \in \mathbf{Z}[\alpha]$, $\beta \neq 0$, then $l_{\mathfrak{p}}(\beta) > 0$ if and only if $\beta \in \mathfrak{p}$;
(c)   for each $\beta \in K^*$ one has $l_{\mathfrak{p}}(\beta) = 0$ for all but finitely many $\mathfrak{p}$, and

$$\prod_{\mathfrak{p}} (\mathfrak{N}\mathfrak{p})^{l_{\mathfrak{p}}(\beta)} = |N(\beta)|,$$

where $\mathfrak{p}$ ranges over the set of all primes of $\mathbf{Z}[\alpha]$.

If $\mathbf{Z}[\alpha] = \mathcal{O}$, it suffices to take $l_{\mathfrak{p}}(x)$ equal to the exponent to which $\mathfrak{p}$ appears in the prime ideal factorization of the ideal $x\mathcal{O}$. The proof of 5.4 for the general case is given in Section 7. It does not use algebraic number theory, but depends on the Jordan-Hölder theorem.

Corollary 5.5. *Let a and b be coprime integers and let $\mathfrak{p}$ be a prime of $\mathbf{Z}[\alpha]$. If $\mathfrak{p}$ is not a first degree prime, then $l_{\mathfrak{p}}(a + b\alpha) = 0$. If $\mathfrak{p}$ is a first degree prime, corresponding to a pair p, r, then $l_{\mathfrak{p}}(a + b\alpha) = e_{p,r}(a + b\alpha)$.*

Proof. Let $\mathfrak{p}$ be a prime of $\mathbf{Z}[\alpha]$ with $l_{\mathfrak{p}}(a + b\alpha) > 0$, and let p be the prime number contained in $\mathfrak{p}$. By 5.4(b), the element $a + b\alpha$ maps to 0 under the map $\mathbf{Z}[\alpha] \to \mathbf{Z}[\alpha]/\mathfrak{p}$. If p divides b, then $b\alpha$ also maps to 0, so the same is true for a, and therefore p divides a; this contradicts that $\gcd(a, b) = 1$. It follows that b maps to a non-zero element of $\mathbf{Z}[\alpha]/\mathfrak{p}$. Denote by b' the inverse of the image of b; it belongs to the prime field $\mathbf{F}_p$ of $\mathbf{Z}[\alpha]$. Since $a + b\alpha$ maps to 0, the element α maps to $-ab'$, which belongs to $\mathbf{F}_p$. Therefore all of $\mathbf{Z}[\alpha]$ maps to $\mathbf{F}_p$, which proves that $\mathfrak{p}$ is a first degree prime. This implies the first assertion of 5.5. If $\mathfrak{p}$ corresponds to p, r, then r is determined by $a + br \equiv 0 \bmod p$. This shows that $\mathfrak{p}$ is the unique prime of $\mathbf{Z}[\alpha]$ containing p and $a + b\alpha$. Now the last statement of 5.5 follows if one compares the power of p on both sides of 5.4(c). This proves 5.5.

We can now prove Proposition 5.3. Let $\prod_{(a,b) \in S} (a + b\alpha) = \gamma^2$, and let $\mathfrak{p}$ be the first degree prime corresponding to p, r. Since $l_{\mathfrak{p}}$ is a homomorphism, we have

$$\sum_{(a,b)\in S} e_{p,r}(a+b\alpha) = \sum_{(a,b)\in S} l_{\mathfrak{p}}(a+b\alpha) = l_{\mathfrak{p}}\Big(\prod_{(a,b)\in S}(a+b\alpha)\Big)$$

$$= l_{\mathfrak{p}}(\gamma^2) = 2l_{\mathfrak{p}}(\gamma) \equiv 0 \bmod 2.$$

This proves 5.3.

6. FOUR OBSTRUCTIONS

We retain the previous notation and remind the reader that we are trying to find a square in $\mathbf{Z}[\alpha]$ by finding a non-empty subset S of

$$T_2 = \{(a,b) \in U : a+b\alpha \text{ is } y\text{-smooth}\}$$

such that the product, over all $(a,b) \in S$, of $a+b\alpha$ is a perfect square in $\mathbf{Z}[\alpha]$.

The first degree primes $\mathfrak{p}$ of $\mathbf{Z}[\alpha]$ of norm at most y comprise the algebraic part of the factor base. Suppose there are exactly B' of them. (We expect B' to be close to $\pi(y)$—see Remark 5.2.) If $\#T_2 > B'$ the linear algebra described in Section 4 can be modified to give us a non-empty set $S \subset T_2$ such that

(6.1) $$\sum_{(a,b)\in S} l_{\mathfrak{p}}(a+b\alpha) \equiv 0 \bmod 2 \qquad \text{for all } \mathfrak{p}.$$

This is weaker than we want. In fact there are four obstructions that may prevent a set S that satisfies (6.1) from satisfying (4.2):

(6.2) The ideal $\prod_{(a,b)\in S}(a+b\alpha)\mathcal{O}$ of $\mathcal{O}$ may not be the square of an ideal, since we work with primes of $\mathbf{Z}[\alpha]$ rather than with primes of $\mathcal{O}$.

(6.3) Even if $\prod_{(a,b)\in S}(a+b\alpha)\mathcal{O} = \mathfrak{a}^2$ for some ideal $\mathfrak{a}$ of $\mathcal{O}$, the ideal $\mathfrak{a}$ need not be principal.

(6.4) Even if $\prod_{(a,b)\in S}(a+b\alpha)\mathcal{O} = \gamma^2\mathcal{O}$ for some $\gamma \in \mathcal{O}$, it is not necessary that $\prod_{(a,b)\in S}(a+b\alpha) = \gamma^2$.

(6.5) Even if $\prod_{(a,b)\in S}(a+b\alpha) = \gamma^2$ for some $\gamma \in \mathcal{O}$, we need not have $\gamma \in \mathbf{Z}[\alpha]$.

We remark that if $\mathbf{Z}[\alpha] = \mathcal{O}$ then the obstructions (6.2) and (6.5) cannot occur. Further, if $\mathcal{O}$ has class number one, and is hence a principal ideal domain, then obstruction (6.3) cannot occur. Finally, if $\mathcal{O}$ is a principal ideal domain and we have an explicit basis for the unit group of $\mathcal{O}$ then we can handle the obstruction (6.4) by linear algebra by including a system of generating units in our factor base. However, in general we cannot make any of these assumptions.

First we note that the fourth obstruction can be dealt with very easily. Namely, if

$$\prod_{(a,b)\in S}(a+b\alpha) = \gamma^2$$

with $\gamma \in K$, then $\gamma \in \mathcal{O}$ and $\gamma f'(\alpha) \in \mathbf{Z}[\alpha]$ (see [40, Proposition 3-7-14]), so

$$(6.6) \qquad f'(\alpha)^2 \cdot \prod_{(a,b)\in S} (a + b\alpha) \quad \text{is the square of an element of } \mathbf{Z}[\alpha].$$

Thus we may replace (4.2) with (6.6) in our factoring algorithm if we also multiply (4.1) by $f'(m)^2$. Indeed, if f and m are chosen by the base m algorithm then $1 < f'(m) < n$ so that we can assume that $\gcd(f'(m), n) = 1$ (since otherwise n would be factored); thus multiplying (4.1) by $f'(m)^2$ will not affect our chance of factoring n.

We could have dealt with the first obstruction by working with the primes $\mathfrak{p}$ of $\mathcal{O}$ rather than those of $\mathbf{Z}[\alpha]$. There is an efficient algorithm for constructing the functions $l_{\mathfrak{p}}$ for those primes, given in [6] (cf. [27, Theorem 4.9]). In practice—or perhaps in the application of the number field sieve to the discrete logarithm problem in a finite field as in [15; 35]—it may be better to use the algorithm from [6]. However, it turns out that the techniques we have to use anyway, in order to cope with obstructions (6.3) and (6.4), also can be used to get around the difference between $\mathbf{Z}[\alpha]$ and $\mathcal{O}$. Thus for simplicity we do not use the algorithm of [6] in what follows.

In Section 8 we describe how to deal with (6.2), (6.3) and (6.4); in the remainder of this section we show that these obstructions are, in a suitable sense, "small" obstructions.

Denote by V the multiplicative group of those $\beta \in K^*$ with the property that $l_{\mathfrak{p}}(\beta) \equiv 0 \bmod 2$ for all primes $\mathfrak{p}$ of $\mathbf{Z}[\alpha]$. Since each $l_{\mathfrak{p}}$ is a group homomorphism, we have $K^{*2} \subset V$. The quotient V/K^{*2} is a vector space over $\mathbf{F}_2$ in a natural way. We can readily produce elements of V but would like elements of K^{*2}; we can measure our obstructions precisely by bounding the dimension of the quotient.

Theorem 6.7. *Let n, d be integers with $d \geq 2$ and $n > d^{2d^2}$, and let m, f be as produced by the base m algorithm in Section 3. Let $K = \mathbf{Q}(\alpha)$ be as in Section 5, and V as defined above. Then we have $\dim_{\mathbf{F}_2} V/K^{*2} < (\log n)/\log 2$.*

Note that this is equivalent to $[V : K^{*2}] < n$. Note also that the bound $n > d^{2d^2}$ supersedes the bound $n > 2^{d^2}$ required in Section 3.

We prove 6.7. Define

$$W = \{\gamma \in K^* : \gamma\mathcal{O} = \mathfrak{a}^2 \text{ for some fractional } \mathcal{O}\text{-ideal } \mathfrak{a}\}.$$

In Section 7 we shall prove that

$$(6.8) \qquad V \supset W, \qquad [V : W] \leq [\mathcal{O} : \mathbf{Z}[\alpha]].$$

Let $Y = \mathcal{O}^* K^{*2}$, where $\mathcal{O}^*$ denotes the group of units of $\mathcal{O}$. Note that the chain of subgroups

$$V \supset W \supset Y \supset K^{*2}$$

corresponds exactly to the first three obstructions.

The index of W in V is bounded by (6.8). Next we consider W/Y. If $\gamma \in W$, then $\gamma\mathcal{O} = \mathfrak{a}^2$ for some fractional $\mathcal{O}$-ideal $\mathfrak{a}$, and the map that sends γ to the ideal class of $\mathfrak{a}$ in the ideal class group of $\mathcal{O}$ clearly has Y as its kernel. We conclude that if h is the order of the class group of K, then

$$[W : Y] \leq h.$$

Finally, Y/K^{*2} is isomorphic to $\mathcal{O}^*/\mathcal{O}^{*2}$, of which the $\mathbf{F}_2$-dimension is equal to the rank of the unit group $\mathcal{O}^*$ plus one (accounting for the roots of unity). Thus from Dirichlet's unit theorem we have

$$[Y : K^{*2}] = 2^{d-s},$$

where s is one-half the number of non-real embeddings of K in the field of complex numbers.

Combining the estimates, we find that

$$[V : K^{*2}] \leq [\mathcal{O} : \mathbf{Z}[\alpha]] \cdot h \cdot 2^{d-s}.$$

Let Δ_K denote the discriminant of K. From [27, Theorem 6.5, Remark] we have that

$$h \leq M \cdot \frac{(d - 1 + \log M)^{d-1}}{(d-1)!},$$

where $M = (d!/d^d)(4/\pi)^s \sqrt{|\Delta_K|}$ is the Minkowski constant of K. Let Δ denote, as in 3.3, the discriminant of f. Then we have

$$M \leq \sqrt{|\Delta_K|} \leq \sqrt{|\Delta_K|} \cdot [\mathcal{O} : \mathbf{Z}[\alpha]] = \sqrt{|\Delta|} < d^d n^{1-3/(2d)}.$$

The equality follows from [9, Chapter I, Section 3, Proposition 4(i) and Section 4, Proposition 6(ii)], and the last inequality is Lemma 3.3. From $d \geq 2$ and $n > d^{2d^2}$ one deduces with a little work that

$$d - 1 + d\log d < \frac{3}{2d}\log n, \qquad 2d \cdot (2\log n)^{d-1} < n^{3/(2d)}.$$

Combining all this, we obtain

$$
\begin{aligned}
[V : K^{*2}] &\leq [\mathcal{O} : \mathbf{Z}[\alpha]] \cdot \frac{d!}{d^d}\left(\frac{4}{\pi}\right)^s \sqrt{|\Delta_K|} \cdot \frac{(d - 1 + \log\sqrt{|\Delta|})^{d-1}}{(d-1)!} \cdot 2^{d-s} \\
&= \frac{\sqrt{|\Delta|}}{d^{d-1}} \cdot 2^d \cdot (d - 1 + \log\sqrt{|\Delta|})^{d-1} \cdot \left(\frac{2}{\pi}\right)^s \\
&< n^{1-3/(2d)} \cdot d \cdot 2^d \cdot \left(d - 1 + d\log d + \left(1 - \frac{3}{2d}\right)\log n\right)^{d-1} \\
&< n^{1-3/(2d)} \cdot 2d \cdot (2\log n)^{d-1} < n,
\end{aligned}
$$

as required. This proves Theorem 6.7.

7. Algebraic Interlude

This section is devoted to the proof of 5.4 and (6.8); it can be skipped by the reader who is willing to take those assertions for granted. Our fundamental tool is the Jordan-Hölder theorem. One can also prove these results using some of the machinery of commutative algebra; for instance, some of the facts proved here can be extracted, with some work, from Appendices A1–3 in [13].

We denote by K an algebraic number field, i.e., a finite field extension of the field $\mathbf{Q}$ of rational numbers, and by K^* its multiplicative group. We let A be an *order* in K, i.e., a subring (with 1) of the ring of integers $\mathcal{O}$ of K with the property that the index of the additive group of A in that of $\mathcal{O}$ is finite. The case of interest in 5.4 is $A = \mathbf{Z}[\alpha]$. In $\mathcal{O}$ one has unique factorization of ideals into prime ideals; in the present section we develop for A a substitute for this unique factorization that meets the needs of the number field sieve.

Let $N: K \to \mathbf{Q}$ be the norm map. For each $x \in K$, the norm $N(x)$ of x is defined to be the determinant of the $\mathbf{Q}$-linear map $K \to K$ that sends each $y \in K$ to xy. It follows that for each non-zero element $x \in A$ we have $\#A/xA = |N(x)|$. This implies that $A/\mathfrak{a}$ is *finite* for each non-zero ideal $\mathfrak{a}$ of A. The cardinality of $A/\mathfrak{a}$ is called the *norm* of $\mathfrak{a}$, denoted $\mathfrak{N}\mathfrak{a}$. In particular, if $\mathfrak{p}$ is a non-zero prime ideal of A, then $A/\mathfrak{p}$ is a finite integral domain, and therefore a field. Hence every such $\mathfrak{p}$ is a maximal ideal of A and contains a unique prime number p; the *degree* of $\mathfrak{p}$ is the degree of $A/\mathfrak{p}$ as a field extension of its prime field $\mathbf{F}_p$. In the sequel, by a "prime of A" we will mean a non-zero prime ideal of A.

The following result clearly contains 5.4 as the special case $A = \mathbf{Z}[\alpha]$.

Proposition 7.1. *There exists, for each prime $\mathfrak{p}$ of A, a group homomorphism $l_\mathfrak{p}: K^* \to \mathbf{Z}$, such that the following hold:*
(a) $l_\mathfrak{p}(x) \geq 0$ *for all* $x \in A$, $x \neq 0$;
(b) *if x is a non-zero element of A, then $l_\mathfrak{p}(x) > 0$ if and only if $x \in \mathfrak{p}$;*
(c) *for each $x \in K^*$ one has $l_\mathfrak{p}(x) = 0$ for all but finitely many $\mathfrak{p}$, and*

$$\prod_\mathfrak{p} (\mathfrak{N}\mathfrak{p})^{l_\mathfrak{p}(x)} = |N(x)|,$$

where $\mathfrak{p}$ ranges over the set of all primes of A.

Proof. First we construct the functions $l_\mathfrak{p}$. Let $\mathfrak{p}$ be a prime of A and let $x \in A$, $x \neq 0$. Since xA is of finite index in A, there is a finite chain

$$A = \mathfrak{a}_0 \supset \mathfrak{a}_1 \supset \mathfrak{a}_2 \supset \cdots \supset \mathfrak{a}_{t-1} \supset \mathfrak{a}_t = xA$$

of distinct ideals of A that cannot be refined, in the sense that there is no ideal properly between $\mathfrak{a}_{i-1}$ and $\mathfrak{a}_i$, for $1 \leq i \leq t$. We now define $l_\mathfrak{p}(x)$ to be the number of $i \in \{1, 2, \ldots, t\}$ for which the A-modules $\mathfrak{a}_{i-1}/\mathfrak{a}_i$ and $A/\mathfrak{p}$ are isomorphic. (We shall see in a moment that for every i there exists a unique such $\mathfrak{p}$.) It follows from the Jordan-Hölder theorem (see [37, Section 51]) that $l_\mathfrak{p}(x)$ does not depend on the choice of the chain of ideals $\mathfrak{a}_i$. (In terms of commutative algebra, $l_\mathfrak{p}(x)$ is the length of the module $A_\mathfrak{p}/xA_\mathfrak{p}$ over the local ring $A_\mathfrak{p}$.)

If x, y are non-zero elements of A, then a chain $\mathfrak{a}_0, \mathfrak{a}_1, \ldots, \mathfrak{a}_t$ as above can be combined with a similar chain $\mathfrak{b}_0, \mathfrak{b}_1, \ldots, \mathfrak{b}_u$ for y into a chain $\mathfrak{a}_0, \mathfrak{a}_1, \ldots,$ $\mathfrak{a}_t = x\mathfrak{b}_0, x\mathfrak{b}_1, \ldots, x\mathfrak{b}_u$ for xy. This proves that we have $l_\mathfrak{p}(xy) = l_\mathfrak{p}(x) + l_\mathfrak{p}(y)$. Therefore we can extend the map $l_\mathfrak{p}$ to a well-defined group homomorphism $K^* \to \mathbf{Z}$ by putting $l_\mathfrak{p}(x/z) = l_\mathfrak{p}(x) - l_\mathfrak{p}(z)$ for any two non-zero elements x, $z \in A$. This completes the construction of the homomorphisms $l_\mathfrak{p}$. It is clear that (a) holds.

To prove the "if" part of (b), it suffices to observe that one can take $\mathfrak{a}_1 = \mathfrak{p}$ if $x \in \mathfrak{p}$. For the "only if" part, suppose that $x \notin \mathfrak{p}$. Since $\mathfrak{p}$ is maximal, the ideal $xA + \mathfrak{p}$ equals A, so $xy + z = 1$ for certain $y \in A$, $z \in \mathfrak{p}$. Then $z \equiv 1 \bmod xA$, so multiplication by z induces the identity map $A/xA \to A/xA$. Hence $z \cdot (\mathfrak{a}_{i-1}/\mathfrak{a}_i) = \mathfrak{a}_{i-1}/\mathfrak{a}_i$, which by $z \in \mathfrak{p}$ implies that $\mathfrak{a}_{i-1}/\mathfrak{a}_i$ cannot be isomorphic to $A/\mathfrak{p}$.

It suffices to prove (c) in the case that $x \in A$. Let the $\mathfrak{a}_i$ be as above, so that

$$|N(x)| = \#A/xA = \prod_{i=1}^{t} \#\mathfrak{a}_{i-1}/\mathfrak{a}_i.$$

Thus to prove (c) it suffices to show that for each i there is a unique prime $\mathfrak{p}$ of A with $\mathfrak{a}_{i-1}/\mathfrak{a}_i \cong A/\mathfrak{p}$. Let $y \in \mathfrak{a}_{i-1}$, $y \notin \mathfrak{a}_i$. Since there is no ideal properly in between $\mathfrak{a}_i$ and $\mathfrak{a}_{i-1}$, we have $yA + \mathfrak{a}_i = \mathfrak{a}_{i-1}$, so multiplication by y induces a surjective map $A \twoheadrightarrow \mathfrak{a}_{i-1}/\mathfrak{a}_i$. Therefore $A/\mathfrak{p} \cong \mathfrak{a}_{i-1}/\mathfrak{a}_i$ for some ideal $\mathfrak{p}$, and since this module has no non-trivial submodules the ideal $\mathfrak{p}$ is maximal. Also, $\mathfrak{p}$ is the annihilator of the A-module $\mathfrak{a}_{i-1}/\mathfrak{a}_i$, so it is uniquely determined. This proves 7.1.

Remark. We remark that the functions $l_\mathfrak{p}$ are uniquely determined by the properties listed in 7.1. To prove this, let $l'_\mathfrak{p}$, for each prime $\mathfrak{p}$ of A, be a homomorphism $K^* \to \mathbf{Z}$, such that (a), (b), (c) hold with $l'_\mathfrak{p}$ instead of $l_\mathfrak{p}$. Let $\mathfrak{p}$ be a prime of A, and p the prime number with $p \in \mathfrak{p}$. Let $x \in A$, $x \neq 0$. To prove that $l'_\mathfrak{p}(x)$ is uniquely determined we proceed as follows. From the definition of $l_\mathfrak{p}$ we see that $\mathfrak{p}^m \mathfrak{b} \subset pxA$, where

$$m = l_\mathfrak{p}(px), \qquad \mathfrak{b} = \prod_{\mathfrak{q} \neq \mathfrak{p}} \mathfrak{q}^{l_\mathfrak{q}(px)}.$$

From $\mathfrak{p}^m + \mathfrak{b} = A$ and the Chinese remainder theorem it follows that there exist y, $z \in A$ with $y \equiv x \bmod \mathfrak{p}^m$, $y \equiv 1 \bmod \mathfrak{b}$, $z \equiv 1 \bmod \mathfrak{p}^m$, $z \equiv x \bmod \mathfrak{b}$. Then $yz \equiv x \bmod pxA$, so $yz = wx$ with $w \equiv 1 \bmod pA$. From z, $w \notin \mathfrak{p}$ one obtains $l'_\mathfrak{p}(x) = l'_\mathfrak{p}(y)$. We have $y \notin \mathfrak{p}'$ for any $\mathfrak{p}' \neq \mathfrak{p}$ that is of p-power norm, since each such $\mathfrak{p}'$ divides $\mathfrak{b}$. Hence $l'_\mathfrak{p}(y)$ can be read off from (c). This proves the uniqueness.

From the uniqueness it follows that in the case $A = \mathcal{O}$ the functions $l_\mathfrak{p}$ coincide with the normalized exponential valuations corresponding to the primes of $\mathcal{O}$; in other words, $l_\mathfrak{p}(x)$ is the exponent of the exact power of $\mathfrak{p}$ dividing the ideal $x\mathcal{O}$. One can also see this by writing the ideal $x\mathcal{O}$ as a product of prime ideals, $x\mathcal{O} = \mathfrak{p}_1\mathfrak{p}_2 \cdots \mathfrak{p}_t$, and choosing $\mathfrak{a}_i = \mathfrak{p}_1\mathfrak{p}_2 \cdots \mathfrak{p}_i$.

We now turn to the proof of (6.8). In the rest of this section A and B denote orders in K with $A \subset B$; for (6.8), we shall take $A = \mathbf{Z}[\alpha]$, $B = \mathcal{O}$. If $\mathfrak{q}$ is a prime of B, then $\mathfrak{p} = \mathfrak{q} \cap A$ is a prime of A. In this case we say that $\mathfrak{q}$ lies over $\mathfrak{p}$, notation: $\mathfrak{q}|\mathfrak{p}$. If $\mathfrak{q}$ lies over $\mathfrak{p}$, then the finite field $B/\mathfrak{q}$ is a field extension of $A/\mathfrak{p}$, and we denote the degree of this field extension by $f(\mathfrak{q}/\mathfrak{p})$. In order to avoid confusion we shall write $l_{\mathfrak{p},A}$ for what we denoted by $l_\mathfrak{p}$ above.

Proposition 7.2. *Let $\mathfrak{p}$ be a prime of A. Then we have*

$$l_{\mathfrak{p},A}(x) = \sum_{\mathfrak{q}|\mathfrak{p}} f(\mathfrak{q}/\mathfrak{p}) l_{\mathfrak{q},B}(x)$$

for each $x \in K^*$, the sum ranging over the primes $\mathfrak{q}$ of B that lie over $\mathfrak{p}$.

Proof. It is convenient, in this proof, to introduce the following notation. If M is a finite A-module, then we let $l_{\mathfrak{p},A}(M)$ be the number of composition factors of M that are isomorphic to $A/\mathfrak{p}$. With this notation, we have $l_{\mathfrak{p},A}(x) = l_{\mathfrak{p},A}(A/xA)$ for every non-zero element $x \in A$. Note that $l_{\mathfrak{p},A}(M) = l_{\mathfrak{p},A}(L) + l_{\mathfrak{p},A}(M/L)$ whenever L is a submodule of M.

It clearly suffices to prove the formula in 7.2 for $x \in A$. Multiplication by x shows that the A-modules B/A and xB/xA are isomorphic, so $l_{\mathfrak{p},A}(B/A) = l_{\mathfrak{p},A}(xB/xA)$. Therefore we have

$$l_{\mathfrak{p},A}(x) = l_{\mathfrak{p},A}(A/xA) = l_{\mathfrak{p},A}(B/xA) - l_{\mathfrak{p},A}(B/A)$$
$$= l_{\mathfrak{p},A}(B/xA) - l_{\mathfrak{p},A}(xB/xA) = l_{\mathfrak{p},A}(B/xB).$$

Hence the formula in 7.2 is equivalent to the statement that for $M = B/xB$ we have

$$l_{\mathfrak{p},A}(M) = \sum_{\mathfrak{q}|\mathfrak{p}} f(\mathfrak{q}/\mathfrak{p}) l_{\mathfrak{q},B}(M).$$

We prove this formula for any finite B-module M. Choosing a composition series for M we immediately reduce to the case that M is a *simple* B-module, which means that M has exactly two B-submodules ($\{0\}$ and itself). In that case $M \cong B/\mathfrak{q}'$ for some prime $\mathfrak{q}'$ of B, and $l_{\mathfrak{q},B}(M)$ equals 1 or 0 according as $\mathfrak{q} = \mathfrak{q}'$ or $\mathfrak{q} \neq \mathfrak{q}'$. Let $\mathfrak{p}' = \mathfrak{q}' \cap A$. As an A-module, $M = B/\mathfrak{q}'$ is a direct sum of $f(\mathfrak{q}'/\mathfrak{p}')$ copies of $A/\mathfrak{p}'$, so that $l_{\mathfrak{p},A}(M)$ equals $f(\mathfrak{q}'/\mathfrak{p}')$ or 0 according as $\mathfrak{p} = \mathfrak{p}'$ or $\mathfrak{p} \neq \mathfrak{p}'$. Thus the above formula follows by inspection. This proves 7.2.

Note that it follows from 7.2 that for each $\mathfrak{p}$ the set of primes $\mathfrak{q}$ of B lying over $\mathfrak{p}$ is finite and non-empty. We now prove that for all but finitely many $\mathfrak{p}$ it is true that there is exactly one $\mathfrak{q}$ lying over $\mathfrak{p}$, and that it satisfies $f(\mathfrak{q}/\mathfrak{p}) = 1$.

Proposition 7.3. *For all but finitely many primes $\mathfrak{p}$ of A we have $\sum_{\mathfrak{q}|\mathfrak{p}} f(\mathfrak{q}/\mathfrak{p})$ $= 1$. In addition, the integer*

$$\prod_{\mathfrak{p}} (\mathfrak{N}\mathfrak{p})^{-1 + \sum_{\mathfrak{q}|\mathfrak{p}} f(\mathfrak{q}/\mathfrak{p})},$$

with $\mathfrak{p}$ ranging over all primes of A, divides the index $[B : A]$ of A in B.

Proof. Let T be any finite set of primes of A, and let U be the set of primes of B lying over the primes in T. Let the A-ideal $\mathfrak{a}$ be the intersection of the primes $\mathfrak{p} \in T$, and let the B-ideal $\mathfrak{b}$ be the intersection of the primes $\mathfrak{q} \in U$. Then $\mathfrak{a} = \mathfrak{b} \cap A$, so $A/\mathfrak{a}$ is a subring of $B/\mathfrak{b}$, and the index of A in B is divisible by the index of $A/\mathfrak{a}$ in $B/\mathfrak{b}$. By the Chinese remainder theorem, we have $A/\mathfrak{a} \cong \prod_{\mathfrak{p} \in T} A/\mathfrak{p}$, and therefore

$$\#A/\mathfrak{a} = \prod_{\mathfrak{p} \in T} \mathfrak{N}\mathfrak{p}.$$

Likewise we have

$$\#B/\mathfrak{b} = \prod_{\mathfrak{q} \in U} \mathfrak{N}\mathfrak{q} = \prod_{\mathfrak{p} \in T} (\mathfrak{N}\mathfrak{p})^{\sum_{\mathfrak{q}|\mathfrak{p}} f(\mathfrak{q}/\mathfrak{p})}.$$

It follows that $[B : A]$ is divisible by

$$(\#B/\mathfrak{b})/(\#A/\mathfrak{a}) = \prod_{\mathfrak{p} \in T} (\mathfrak{N}\mathfrak{p})^{-1 + \sum_{\mathfrak{q}|\mathfrak{p}} f(\mathfrak{q}/\mathfrak{p})}.$$

Therefore the number of $\mathfrak{p} \in T$ for which $\sum_{\mathfrak{q}|\mathfrak{p}} f(\mathfrak{q}/\mathfrak{p}) \neq 1$ is bounded independently of T, which implies the first assertion of 7.3. Taking for T the set of all $\mathfrak{p}$ with $\sum_{\mathfrak{q}|\mathfrak{p}} f(\mathfrak{q}/\mathfrak{p}) \neq 1$ we obtain the second. This proves 7.3.

In our final result in this section, we write

$$V_A = \{x \in K^* : l_{\mathfrak{p},A}(x) \equiv 0 \bmod 2 \text{ for all primes } \mathfrak{p} \text{ of } A\}.$$

In the notation of (6.8) we clearly have $V_{\mathbf{Z}[\alpha]} = V$ and $V_{\mathcal{O}} = W$. Hence (6.8) is an immediate consequence of the following proposition.

Proposition 7.4. *If $A \subset B$ are orders of K, then $V_B \subset V_A$, and $[V_A : V_B] \leq [B : A]$.*

Proof. The inclusion $V_B \subset V_A$ is clear from 7.2. To bound $[V_A : V_B]$, we choose for each prime $\mathfrak{p}$ of A a set $S_{\mathfrak{p}}$ of primes $\mathfrak{q}$ of B lying over $\mathfrak{p}$, as follows. If $f(\mathfrak{q}/\mathfrak{p})$ is even for each prime $\mathfrak{q}$ of B lying over $\mathfrak{p}$, then we let $S_{\mathfrak{p}}$ be the set of all $\mathfrak{q}$ lying over $\mathfrak{p}$. If there is at least one $\mathfrak{q}$ lying over $\mathfrak{p}$ for which $f(\mathfrak{q}/\mathfrak{p})$ is odd, then we choose one such prime, $\mathfrak{q}_0$ (say), and we let $S_{\mathfrak{p}}$ consist of all primes $\mathfrak{q} \neq \mathfrak{q}_0$ that lie over $\mathfrak{p}$. Since $f(\mathfrak{q}/\mathfrak{p}) \geq 2$ if $f(\mathfrak{q}/\mathfrak{p})$ is even, we have

$$\#S_{\mathfrak{p}} \leq -1 + \sum_{\mathfrak{q}|\mathfrak{p}} f(\mathfrak{q}/\mathfrak{p})$$

for all $\mathfrak{p}$. In particular, $S_{\mathfrak{p}}$ is empty for almost all $\mathfrak{p}$. Let S be the union of the sets $S_{\mathfrak{p}}$, with $\mathfrak{p}$ ranging over the primes of A. We have

$$2^{\#S} \leq \prod_{\mathfrak{p}} (\mathfrak{N}\mathfrak{p})^{\#S_{\mathfrak{p}}} \leq \prod_{\mathfrak{p}} (\mathfrak{N}\mathfrak{p})^{-1 + \sum_{\mathfrak{q}|\mathfrak{p}} f(\mathfrak{q}/\mathfrak{p})} \leq [B : A],$$

by 7.3. Thus to prove 7.4, it suffices to show that the group V_A/V_B embeds in the group $(\mathbf{Z}/2\mathbf{Z})^S$. To do this, map $x \in V_A$ to the element $(l_{\mathfrak{q},B}(x) \bmod 2)_{\mathfrak{q} \in S}$ of $(\mathbf{Z}/2\mathbf{Z})^S$. If x is in the kernel of this map, then $l_{\mathfrak{q},B}(x)$ is even for all $\mathfrak{q} \in S$. Since also all $l_{\mathfrak{p},A}(x)$ are even, it follows from 7.2 and the choice of $S_{\mathfrak{p}}$ that $l_{\mathfrak{q},B}(x)$ is even for all $\mathfrak{q}$, so that $x \in V_B$. This proves 7.4.

8. Quadratic Characters

In this section the notation and hypotheses are as in Sections 4 and 5. We assume in addition that $n > d^{2d^2}$, and that m, f have been produced by the base m method of Section 3.

In our original version of the number field sieve we handled the three obstructions (6.2), (6.3), (6.4) as follows. We dealt with the first obstruction, which is due to the difference between the rings $\mathbf{Z}[\alpha]$ and $\mathcal{O}$, by using the algorithm of [6], as mentioned in Section 6. To overcome the second obstruction, we proposed that the linear algebra on the algebraic side be done over $\mathbf{Z}$ rather than over $\mathbf{F}_2$ (cf. [24, Extended abstract, Section 7]). This allowed the construction of integers $s(a, b)$ for pairs $(a, b) \in T_2$ such that

$$(8.1) \qquad \prod_{(a,b) \in T_2} (a + b\alpha)^{s(a,b)} \mathcal{O} = (1).$$

Thus $\prod (a + b\alpha)^{s(a,b)}$ is a unit. The third obstruction was overcome by means of lattice basis reduction methods on the logarithmic embedding in Euclidean space of the units arising (see [15]). Thus several equations of the form (8.1) could be combined to find integers $s'(a, b)$ such that

$$\prod_{(a,b) \in T_2} (a + b\alpha)^{s'(a,b)} = 1.$$

By then combining these ideas with the sieve on the rational side as discussed in Section 4, we could find integers $s''(a, b)$ for each pair $(a, b) \in T_1 \cap T_2$ such that we have

$$\prod_{(a,b) \in T_1 \cap T_2} (a + bm)^{s''(a,b)} \quad \text{is a square in } \mathbf{Z},$$

$$\prod_{(a,b) \in T_1 \cap T_2} (a + b\alpha)^{s''(a,b)} = 1.$$

These equations could then be used in place of (2.1) and (2.2) to attempt to factor n.

In addition to being inelegant and complicated, the linear algebra step over $\mathbf{Z}$ in the above scenario became a bottleneck in the complexity argument. In fact the heuristic run time of the above version of the number field sieve is $L_n[\frac{1}{3}, 9^{1/3} + o(1)]$ for $n \to \infty$ rather than the bound we advertised above; the latter could be achieved only at the expense of considerable additional complications.

It was at this point that Adleman [1] suggested using quadratic characters to overcome the second and third obstructions. As we shall see this allows the linear algebra on the algebraic side to be done over $\mathbf{F}_2$, greatly simplifying the algorithm. In fact we use this same idea to also overcome the first obstruction.

In order to explain the idea behind "character columns", we start by considering a simpler situation. Suppose that X is a finite set of primes and that $l \in \mathbf{Z}$,

$l \neq 0$, has the property that in the factorization of l into primes, the exponent of each prime not in X is even. Is l a square? The answer of course depends on the sign of l and the exponent of each prime $p \in X$ in the factorization of l. If these quantities are inaccessible for some reason then we can still test l for squareness by the following probabilistic device: if p is a prime number that is not in X and p does not divide $2l$, then test the Legendre symbol $\left(\frac{l}{p}\right)$ to see if it is equal to 1. If the symbol is ever equal to -1 then l is not a square; if the symbol is always equal to 1 for a number of primes p significantly exceeding $\#X$ then we become convinced that l is a square. Specifically, if V_X denotes the multiplicative group of non-zero rational numbers that are squares outside X as above, then $V_X/\mathbf{Q}^{*2}$ is an $\mathbf{F}_2$-vector space of dimension $\#X + 1$. The Legendre symbol corresponding to each "test" prime p is a presumably random linear function on this vector space. Our test for l being a square is ironclad if the characters corresponding to the primes p that we choose span the dual space of $V_X/\mathbf{Q}^{*2}$.

Lemma 8.2. *Let k, r be non-negative integers, and let E be a k-dimensional $\mathbf{F}_2$-vector space. Then the probability that $k+r$ elements that are independently drawn from E, with the uniform distribution, form a spanning set for E is at least $1 - 2^{-r}$.*

Proof. For any hyperplane H of E, the probability that each of the $k+r$ vectors lies in H is 2^{-k-r}. Since each hyperplane is the kernel of a uniquely determined non-zero linear function $E \to \mathbf{F}_2$, the number of hyperplanes of E is $2^k - 1$. Thus the probability that the $k+r$ vectors all lie in *some* hyperplane is at most

$$(2^k - 1)2^{-k-r} < 2^{-r}.$$

However, the $k+r$ vectors do not span E if and only if they lie in some hyperplane. Thus the lemma follows.

Remark. If one picks random elements of E, independently, and from the uniform distribution, until one has a set of generators, then the expectation of the number of elements drawn is equal to $k + \sum_{i=1}^{k}(2^i - 1)^{-1}$. For $k \to \infty$, the sum tends to a limit c where $c \doteq 1.606695$. Thus for any k, the expectation is less than $k + 2$.

If we had some method of choosing Legendre characters that in the above scenario corresponds to choosing elements of the dual space of $V_X/\mathbf{Q}^{*2}$ independently and from the uniform distribution, then we could develop a virtually certain test for squareness for the integer l. In what follows, we replace $\mathbf{Z}$ with $\mathbf{Z}[\alpha]$ and make the heuristic assumption that choosing Legendre characters corresponding to small primes outside the factor base suffices for a squareness test.

The following result shows how Legendre symbols provide us with a necessary condition for a product of elements $a + b\alpha$ to be a square. The set $R(q)$ is as defined after (5.1).

Proposition 8.3. *Let S be a finite set of coprime integer pairs (a, b) with the property that $\prod_{(a,b) \in S} (a + b\alpha)$ is the square of an element of K. Further let q*

be an *odd prime number and* $s \in R(q)$, *such that*

$$a + bs \not\equiv 0 \bmod q \qquad \text{for each } (a, b) \in S,$$
$$f'(s) \not\equiv 0 \bmod q.$$

Then we have

$$\prod_{(a,b)\in S} \left(\frac{a + bs}{q} \right) = 1.$$

Proof. Let $\mathbf{Z}[\alpha] \to \mathbf{F}_q$ be the ring homomorphism mapping α to $s \bmod q$, and let $\mathfrak{q}$ be its kernel; this is the first degree prime corresponding to q, s. Define the map $\chi_{\mathfrak{q}} : \mathbf{Z}[\alpha] - \mathfrak{q} \to \{\pm 1\}$ to be the composition of $\mathbf{Z}[\alpha] - \mathfrak{q} \to \mathbf{F}_q - \{0\}$ with the Legendre symbol $\mathbf{F}_q - \{0\} \to \{\pm 1\}$. Clearly, we have $\chi_{\mathfrak{q}}(a + b\alpha) = \left(\frac{a+bs}{q} \right)$.

As we saw in (6.6), we have

$$f'(\alpha)^2 \cdot \prod_{(a,b)\in S} (a + b\alpha) = \delta^2$$

for some $\delta \in \mathbf{Z}[\alpha]$. By hypothesis, the factors on the left are not in $\mathfrak{q}$, so we have $\delta \notin \mathfrak{q}$. The proposition follows if we apply $\chi_{\mathfrak{q}}$ to the equation.

As with 5.3, it is really the converse to 8.3 that we are interested in, and in this case it does hold: if an element $\beta \in \mathbf{Z}[\alpha] - \{0\}$ satisfies $\chi_{\mathfrak{q}}(\beta) = 1$ for all first degree primes $\mathfrak{q}$ with $2\beta \notin \mathfrak{q}$, or even for all such $\mathfrak{q}$ with finitely many exceptions, then β is a square in K.

In the actual algorithm, we use both the functions $e_{p,r}$ and the Legendre symbols to produce the square that we need, as follows. Let $T = T_1 \cap T_2$, so that

$$T = \{(a, b) : \gcd(a, b) = 1, |a| \leq u, 0 < b \leq u, (a + bm)N(a + b\alpha) \text{ is } y\text{-smooth}\}.$$

Define

$$B = \pi(y),$$
$$B' = \#\{(p, r) : p \text{ is a prime number}, p \leq y, r \in R(p)\},$$
$$B'' = [3(\log n)/\log 2].$$

We define the factor base on the rational side to be the set of all prime numbers up to y, call them $p_1, p_2, \ldots, p_B$. Define the factor base on the algebraic side to be the set of pairs $(p_1, r_1), (p_2, r_2), \ldots, (p_{B'}, r_{B'})$ as in the definition of B'. Let $(q_1, s_1), (q_2, s_2), \ldots, (q_{B''}, s_{B''})$ be the first B'' pairs consisting of a prime number $q > y$ and an integer $s \in R(q)$ with $f'(s) \not\equiv 0 \bmod q$, ordered by increasing q.

We now define a map e from T to $\mathbf{F}_2^{1+B+B'+B''}$. Say $(a, b) \in T$. The first coordinate of $e(a, b)$ is determined by the sign of $a + bm$; it is 0 if $a + bm > 0$ and 1 if $a + bm < 0$ (we cannot have $a + bm = 0$ if $m > u$, which will be the case with our choice of parameters; see Section 11). The next B coordinates are given by $\operatorname{ord}_p(a + bm) \bmod 2$ as p runs over $p_1, p_2, \ldots, p_B$. The next B' coordinates are given by $e_{p,r}(a + b\alpha) \bmod 2$ as (p, r) runs over $(p_1, r_1), (p_2, r_2), \ldots, (p_{B'}, r_{B'})$. The last B'' coordinates of $e(a, b)$ are determined by $\left(\frac{a+bs}{q} \right)$ as (q, s) runs over

(q_1, s_1), (q_2, s_2), ..., $(q_{B''}, s_{B''})$. For a particular (q, s) it is 0 if $\left(\frac{a+bs}{q}\right) = 1$ and 1 if $\left(\frac{a+bs}{q}\right) = -1$. Note that the reason for the special treatment of the first coordinate and the last B'' coordinates is to turn a multiplicative structure into an additive structure.

If $\#T > 1 + B + B' + B''$ then the vectors $e(a, b)$ for $(a, b) \in T$ are linearly dependent. Thus there is a non-empty subset S of T such that $\sum_{(a,b) \in S} e(a, b)$ is the zero vector in $\mathbf{F}_2^{1+B+B'+B''}$. It is clear that such a set satisfies (4.1), and we conjecture that it satisfies (6.6) as well.

To support this conjecture, we make the following remarks. Let V be the subgroup of K^* defined before Theorem 6.7. If $\mathfrak{q}$ is any first degree prime of $\mathbf{Z}[\alpha]$ with $f'(\alpha) \notin \mathfrak{q}$, then the function $\chi_\mathfrak{q}$ defined in the proof of 8.3 induces a group homomorphism $V/K^{*2} \to \{\pm 1\}$, again to be denoted by $\chi_\mathfrak{q}$; namely, one can show that any $\beta \in V$ can be written as $\beta = \beta_1 \beta_2^2$, with $\beta_1 \in \mathbf{Z}[\alpha] - \mathfrak{q}$ and $\beta_2 \in K^*$, and that $\chi_\mathfrak{q}(\beta_1)$ is independent of this representation, so that we can put $\chi_\mathfrak{q}(\beta) = \chi_\mathfrak{q}(\beta_1)$. The Čebotarev density theorem (see [19, Chapter VIII, Section 4]) implies that if $\mathfrak{q}$ ranges over all first degree primes of $\mathbf{Z}[\alpha]$ with $f'(\alpha) \notin \mathfrak{q}$, ordered by increasing norm, then the elements $\chi_\mathfrak{q}$ are asymptotically equally distributed over $\mathrm{Hom}(V/K^{*2}, \{\pm 1\})$. This suggests that the B'' functions $\chi_\mathfrak{q}$ that the algorithm employs may be viewed as random homomorphisms $V/K^{*2} \to \{\pm 1\}$, so that Theorem 6.7 and Lemma 8.2 make it overwhelmingly likely that these functions $\chi_\mathfrak{q}$ span $\mathrm{Hom}(V/K^{*2}, \{\pm 1\})$. If they do, then for an element $\beta \in V$ to be a square it would be necessary and sufficient that $\chi_\mathfrak{q}(\beta) = 1$ for each of the B'' primes $\mathfrak{q}$, which would imply the conjecture. A rigorous proof of the conjecture along these lines would require a very strong effective version of the Čebotarev density theorem, which presently appears to be completely out of reach. It may be possible to deduce a weak form of the conjecture—with B'' replaced by a larger value—from the generalized Riemann hypothesis (cf. [2]). In addition, it may be possible to rigorously prove a *random* version of the above, where the B'' primes $\mathfrak{q}$ are independently and uniformly chosen from all the first degree primes of $\mathbf{Z}[\alpha]$ in some reasonable range.

Remark. One can also make use of Legendre symbols that are defined for primes $\mathfrak{q}$ of odd norm that have degree greater than 1. However, there is a certain danger involved in using these primes. For example, if $d = 2$, then the base m method of Section 3 leads to an imaginary quadratic field, and one can show that in that case $\chi_\mathfrak{q}(u) = 1$ for every unit u of $\mathcal{O}$ and every prime $\mathfrak{q}$ of odd norm of degree greater than 1; this means that the quadratic characters associated to such primes are not sufficient to deal with obstruction (6.4). First degree primes do not suffer from this shortcoming.

9. FINDING SQUARE ROOTS

We retain the notation and hypotheses from the last section.

Now that we have produced presumed squares in $\mathbf{Z}$ and $\mathbf{Z}[\alpha]$ we need to find their square roots. In $\mathbf{Z}$ this is easy. If $f'(m)^2 \prod_{(a,b) \in S} (a + bm)$ is a square, then since the prime factorization of each $a + bm$ is known it is an easy matter to compute the square root. We are ultimately only interested in the result mod n, so all of the arithmetic can be done with integers of the size of n.

Next we address the problem of finding the square root in the number field. This is a component of the number field sieve that has no analogue in earlier factoring algorithms, including the special number field sieve. In the known solutions to this problem one cannot work "mod n", as we did in $\mathbf{Z}$, and it is hard to see how one can avoid dealing with numbers of a truly gigantic size. In fact, the methods that we discuss in this section require arithmetic with numbers of about $\sqrt{C}$ digits, where C is the running time of the entire number field sieve (see 9.3 and Section 11). (In all other components of the number field sieve we work with numbers of only $C^{o(1)}$ digits, for $n \to \infty$.) The time needed to find the square root may dominate the running time of the number field sieve, unless one employs techniques depending on the fast Fourier transform. In this section we discuss the problem from a theoretical point of view. Practical experiments conducted by D. J. Bernstein indicate that the method that we shall suggest might actually work in practice. Meanwhile, J.-M. Couveignes [11] discovered a more practical algorithm that does avoid large numbers; it requires the degree d to be odd.

Let $\gamma = f'(\alpha)^2 \prod_{(a,b) \in S} (a + b\alpha)$ be the presumed square in $\mathbf{Z}[\alpha]$. To find its square root, we can first multiply out the product and represent γ as a polynomial in α of degree less than d, and next apply one of the algorithms that have been proposed for factoring polynomials over algebraic number fields (see [38; 39; 18; 21]) to the polynomial $X^2 - \gamma \in K[X]$. It is important to bear in mind that, when all parameters of the number field sieve are chosen optimally, the cardinality of the set S and the coefficients of γ as a polynomial in α are very large (see 9.3 and Section 11). This implies that just *computing* γ is already very time consuming, and factoring $X^2 - \gamma$ even more so. In order to be able to analyze the complexity of this step we consider what the algorithms of [38; 39; 18; 21] come down to in our case.

There is no essential difference between the algorithms proposed in [38; 39; 18; 21] if an odd prime number q is available for which f mod q is irreducible in $\mathbf{F}_q[X]$; so let this now first be assumed. Then $\mathbf{Z}[\alpha]/q\mathbf{Z}[\alpha]$ is isomorphic to $\mathbf{F}_q[X]/(f \bmod q)$, which is a field of cardinality q^d. Hence the ideal $\mathfrak{q} = q\mathbf{Z}[\alpha]$, which consists of all elements $\sum_{i=0}^{d-1} a_i \alpha^i$ for which each of the integer coefficients a_i is divisible by q, is a prime of $\mathbf{Z}[\alpha]$ of degree d. From the irreducibility of f mod q it follows that $f'(\alpha) \notin \mathfrak{q}$, and for each $(a,b) \in S$ we have $a + b\alpha \notin \mathfrak{q}$ since $\gcd(a,b) = 1$. Therefore the product γ of all these elements does not belong to $\mathfrak{q}$ either. Taking the coefficients of γ modulo q, and applying an algorithm for taking square roots in the finite field $\mathbf{Z}[\alpha]/\mathfrak{q}$ (see [20; 17, Section 4.6.2, Exercise 15]), we find an element $\delta_0 \pmod{\mathfrak{q}}$ such that $\delta_0^2 \gamma \equiv 1 \bmod \mathfrak{q}$; this $\delta_0 \bmod \mathfrak{q}$ is unique up to sign. (If one finds, unexpectedly, that $X^2 - \gamma$ is actually irreducible modulo $\mathfrak{q}$, so that δ_0 cannot be found, then γ is not a square in $\mathbf{Z}[\alpha]$, and we have hit upon a counterexample to the conjecture stated in Section 8. In this case more character columns might be tried.) Note that δ_0 is the *inverse* of a square root of γ mod $\mathfrak{q}$; this is in order to avoid divisions in the iteration to follow. Starting from δ_0, we apply a Newton iteration

$$\delta_j \equiv \frac{\delta_{j-1}(3 - \delta_{j-1}^2 \gamma)}{2} \bmod \mathfrak{q}^{2^j}$$

to find δ_1, δ_2, ..., such that $\delta_j^2 \gamma \equiv 1 \bmod \mathfrak{q}^{2^j}$. Notice that working modulo $\mathfrak{q}^{2^j}$ means that the coefficients a_i in the expressions $\sum_i a_i \alpha^i$ are taken modulo q^{2^j}, so that one may take $|a_i| < q^{2^j}/2$. One continues the Newton iteration until q^{2^j} is at least twice as large as an upper bound that one is able to prove for the absolute values of the coefficients of a true square root β of γ in $\mathbf{Z}[\alpha]$. Then β can be calculated from $\beta \equiv \delta_j \gamma \bmod \mathfrak{q}^{2^j}$. If we wish, we can now verify that $\beta^2 = \gamma$, and thus free ourselves from having to rely on the unproved conjecture of Section 8; but in the context of the number field sieve it is more efficient to just assume that $\beta^2 = \gamma$, and to proceed immediately to the calculation of $\varphi(\beta)$ (as in Section 2) in an attempt to factor n.

There are several refinements and modifications that might affect the practical performance of this scheme. For example, one can apply fast multiplication techniques in the iteration; one can go up by powers $\mathfrak{q}^j$ instead of $\mathfrak{q}^{2^j}$ of $\mathfrak{q}$; and one can stop the iteration as soon as the coefficients of $\delta_j \gamma \bmod \mathfrak{q}^{2^j}$ do not change for a few successive values of j. One may also wonder whether there is a method that does not start by multiplying out the product that defines γ.

In the above description we made the assumption that an odd prime number q is available for which $f \bmod q$ is irreducible. One can attempt to find such a prime number q by trying $q = 3, 5, 7, \ldots$ in succession. (Of course, the prime numbers that are norms of first degree primes of $\mathbf{Z}[\alpha]$ can be left out.) For each q, one can test $f \bmod q$ for irreducibility by applying an irreducibility test in $\mathbf{F}_q[X]$ (see [20]). As we shall see below, one may for most n expect to be successful fairly soon. However, there are cases in which not a single prime number q exists for which $f \bmod q$ is irreducible. This occurs, for example, when

$$n = m^4 + 1, \qquad d = 4.$$

The question arises how to proceed when this happens.

One solution of this problem is based on the remark that, in a sense that can be made precise, most monic polynomials f of degree d in $\mathbf{Z}[X]$ have the property that the Galois group of f is the full symmetric group S_d of order $d!$ (see [14]). If f satisfies this condition, then the Čebotarev density theorem implies that the density, inside the set of all prime numbers, of the set of prime numbers q for which $f \bmod q$ is irreducible is equal to the probability that a random permutation of $\{1, 2, \ldots, d\}$ is a single d-cycle (cf. the proof of 9.1 below), which is equal to $1/d$. Since d will be chosen quite small with respect to n (see Section 11), this is fairly large, so that for most values of n we expect that there are many suitable prime numbers q and that it will be easy to find one. It may be possible to make this loose argument perfectly rigorous. If, for whatever reason, a good q is difficult to find, then one has the option of changing f (and hence the number field), for example by adding a polynomial that is divisible by $X - m$ to f, or by choosing a different value of m in the base m algorithm. However, there are situations in which it is very undesirable to change f, for example when f has particularly small coefficients. In that case one may not be able to work with primes q for which $f \bmod q$ is irreducible.

We briefly discuss what one can do if no odd prime number q is available for which $f \bmod q$ is irreducible. The approach of [39] is then to do a similar Newton iteration modulo powers of an odd prime number q. At the start of the iteration, the ideal $q\mathbf{Z}[\alpha]$ is not prime, so that the inverse square root δ_0 of γ ($\bmod q$) is not unique up to sign. Instead, one must take the inverse square root of γ modulo each of the primes $\mathfrak{q}$ containing q, and combine them into an inverse square root modulo $q\mathbf{Z}[\alpha]$; or if q is small, one can try all $(q^d - 1)/2$ non-zero elements of $\mathbf{Z}[\alpha]/q\mathbf{Z}[\alpha]$, up to sign. If there are t primes $\mathfrak{q}$ containing q, then this gives rise to 2^{t-1} different starting values δ_0 for the Newton iteration. If we choose q as indicated below, then we have $t \le d/2$, and it turns out that, with our choice of parameters, a factor $2^{\lceil d/2 \rceil - 1}$ does not greatly affect the running time; so the algorithm of [39] may be feasible for our purposes.

The polynomial time algorithm of [18; 21] does a Newton iteration modulo the powers of a *single* prime $\mathfrak{q}$ containing q. To recover the square root of γ from $\delta_j \gamma$, for large j, one then needs to apply a basis reduction algorithm to the ideal $\mathfrak{q}^{2^j}$. This is, with our choice of parameters, not attractive (see 9.3). Another possibility is the algorithm of [38], but we have not investigated its merits for use in the number field sieve. A final possibility is to make use of the "infinite" prime, as was pointed out to us by V. S. Miller and R. D. Silverman. In this case, one chooses an element of $K = \mathbf{Q}(\alpha)$ that under each embedding σ of K in the field of complex numbers is close to a square root of $\sigma(\gamma)$, and one next applies a Newton iteration in $\mathbf{Q}(\alpha)$, where one works with the coefficients a_i as real numbers that are rounded to rationals. For this algorithm, the number of different starting values to be tried is 2^{d-s-1}, where s is one-half the number of non-real embeddings of K into the field of complex numbers. For each of these methods, the applicability of the refinements mentioned above is to be considered. Which method is the best one for practical purposes remains to be tested.

If one decides to use the algorithm of [39], then the choice of an appropriate prime number q is still important, since the method requires that the algebraic integer γ be coprime to q. This is guaranteed if $f \bmod q$ factors into distinct irreducible non-linear factors. Indeed, if $f \bmod q$ is squarefree, then q is relatively prime to $f'(\alpha)$, and if $f \bmod q$ has no linear factors then there is no first degree prime of norm q, so that by 5.5 each $a + b\alpha$ is coprime to q. One may wonder whether primes q with the properties just mentioned exist. The following result answers this question affirmatively, and in addition it asserts that there are so many of them that in practice it should not be hard to find one.

Proposition 9.1. *Let $f \in \mathbf{Z}[X]$ be an irreducible monic polynomial of degree d, with $d > 1$. Then the density, inside the set of all prime numbers, of the set of prime numbers q for which $f \bmod q$ factors in $\mathbf{F}_q[X]$ into distinct irreducible* *non-linear factors exists and is at least $1/d$.*

Proof. Let G be the Galois group of f over $\mathbf{Q}$, viewed as a permutation group of the set Ω of zeroes of f. For each prime number q that does not divide the discriminant of f, there is a Frobenius element $\sigma_q \in G$, which is well-defined up to conjugacy in G, and which has the property that the degrees of the irreducible factors of f mod q are the same as the lengths of the cycles of the permutation σ_q. Hence, we are interested in those q for which σ_q acts without fixed points on Ω. The Čebotarev density theorem [19, Chapter VIII, Section 4] implies that for every subset $C \subset G$ that is a closed under conjugation by G, the set of prime numbers q for which σ_q belongs to C has a density, and that this density equals $\#C/\#G$. Hence, the proposition follows from the following fact in group theory, which was kindly proved for us by A. M. Cohen (see [7; 3]).

Lemma 9.2. *Let G be a finite group that acts transitively on a finite set Ω, with $\#\Omega = d > 1$. Then there are at least $(\#G)/d$ elements of G that act without fixed points on Ω.*

Proof. We recall that if G acts on a finite set X, then the number of orbits of X under G is given by the formula

$$\frac{1}{\#G} \sum_{\sigma \in G} \#X^{\sigma},$$

where $X^{\sigma} = \{x \in X : \sigma x = x\}$ (see [16, Kapitel V, Satz 13.4]). We first apply this formula to $X = \Omega$, which by hypothesis has one orbit under G. Writing f_i for the number of $\sigma \in G$ that have exactly i fixed points on Ω, we get

$$\sum_{i=0}^{d} i f_i = \#G.$$

Next we apply it to $X = \Omega \times \Omega$, with G acting componentwise. The diagonal is transformed into itself by G, and there are also off-diagonal points, because $d > 1$. Hence X has at least two orbits under G, so that we obtain

$$\sum_{i=0}^{d} i^2 f_i \geq 2\#G.$$

Finally, we have the trivial relation

$$\sum_{i=0}^{d} f_i = \#G.$$

Since the number $i^2 - (d+1)i + d = (i-1)(i-d)$ is non-positive for $1 \leq i \leq d$, and equal to d for $i = 0$, we now find that

$$d f_0 \geq \sum_{i=0}^{d} (i^2 - (d+1)i + d) f_i \geq (2 - (d+1) + d) \cdot \#G = \#G,$$

as desired. This completes the proof of 9.2 and 9.1.

9.3. *Complexity.* The complexity analysis of the square root algorithm that we described in this section is entirely straightforward. As we shall see in Section 11, the parameters u and y will be chosen as functions of n and d such that each of $\log u$ and $\log y$ equals

$$(\tfrac{1}{2} + o(1))\left(d \log d + \sqrt{(d \log d)^2 + 4 \log(n^{1/d}) \log \log(n^{1/d})}\right)$$

for $n \to \infty$, and the running time of other steps in the algorithm will (heuristically) be bounded by $y^{2+o(1)}$. In addition, we shall have $\#T = y^{1+o(1)}$, so the same expression is an upper bound for $\#S$ as well, and it is unlikely that $\#S$ is much smaller. Thus an upper bound for the absolute value of the integers involved in the computation of a square root of γ is $\exp(y^{1+o(1)})$. In these circumstances, the calculation of the square root of γ as described in this section takes time at most $y^{1+o(1)}$ if one employs fast multiplication techniques, and $y^{2+o(1)}$ if one uses traditional algorithms for the arithmetic operations. Thus if one does not use fast multiplication techniques then the running time of the square root algorithm may dominate the running time of the entire number field sieve. If we replace [39] by [18; 21] in the square root algorithm, then one has to perform a basis reduction algorithm, and the running time bounds become $y^{2+o(1)}$ and $y^{3+o(1)}$, with fast and traditional arithmetic respectively; the numbers one works with are bounded by $\exp(y^{1+o(1)})$, as before. Thus it is not attractive to use the methods of [18; 21].

Remark. To make the above algorithm more efficient, we can attempt to replace the element γ of which we take the square root by an element that has smaller coefficients when expressed as a polynomial in α. This can possibly be achieved by means of the following idea, which bears some resemblance to the square root algorithm of [30]. Suppose $S = \{(a_1, b_1), \ldots, (a_s, b_s)\}$, where $\#S = s$. We inductively define two sequences $(\mu_i)_{i=0}^{s}$ and $(\nu_i)_{i=0}^{s}$ of elements of $\mathbf{Z}[\alpha]$. First let $\mu_0 = \nu_0 = 1$. Suppose $1 \le i \le s$ and μ_{i-1}, ν_{i-1} have been defined. If $a_i + b_i \alpha$ divides μ_{i-1} in $\mathbf{Z}[\alpha]$, we let $\mu_i = \mu_{i-1}/(a_i + b_i \alpha)$ and we let $\nu_i = \nu_{i-1}(a_i + b_i \alpha)$. Otherwise, we let $\mu_i = \mu_{i-1}(a_i + b_i \alpha)$ and $\nu_i = \nu_{i-1}$. We have the identity

$$\gamma = f'(\alpha)^2 \prod_{i=1}^{s}(a_i + b_i \alpha) = f'(\alpha)^2 \mu_s \nu_s^2,$$

so that if γ is a square in $\mathbf{Z}[\alpha]$, so is $f'(\alpha)^2 \mu_s$. Thus, instead of taking a square root of γ, it suffices to take a square root of $f'(\alpha)^2 \mu_s$ and to multiply this square root by ν_s. In addition, our factoring algorithm does not need ν_s itself, but only its image $\varphi(\nu_s)$ in $\mathbf{Z}/n\mathbf{Z}$, which one can calculate by only doing arithmetic with integers the size of n.

To test if some non-zero $a + b\alpha$ divides some μ in $\mathbf{Z}[\alpha]$ and compute the quotient if it does, we divide $a + bX$ into f to get $f = (a + bX)g + f(-a/b)$, where $g \in \mathbf{Q}[X]$. Then $a + b\alpha$ divides μ in $\mathbf{Z}[\alpha]$ if and only if $\mu/(a + b\alpha) = -\mu g(\alpha)/f(-a/b)$ belongs to $\mathbf{Z}[\alpha]$.

By using exponent vectors, one can often see very cheaply that $a_i + b_i\alpha$ does not divide μ_{i-1} in $\mathbf{Z}[\alpha]$. Let (p_1, r_1), (p_2, r_2), ..., $(p_{B'}, r_{B'})$ be the factor base on the algebraic side, and for $1 \leq i \leq s$ let $v_i \in \mathbf{Z}^{B'}$ be the integer vector whose coordinates are the numbers $e_{p,r}(a_i + b_i\alpha)$ as (p, r) runs over (p_1, r_1), (p_2, r_2), ..., $(p_{B'}, r_{B'})$. Define the vectors $w_i \in \mathbf{Z}^{B'}$ inductively by $w_0 = 0$, $w_i = w_{i-1} - v_i$ if $a_i + b_i\alpha$ divides μ_{i-1}, and $w_i = w_{i-1} + v_i$ otherwise. From Proposition 7.1 we see that w_i is the exponent vector of μ_i and that it has non-negative coordinates. This gives an easily checked necessary condition for $a_i + b_i\alpha$ to divide μ_{i-1}, namely that $w_{i-1} - v_i$ has non-negative coordinates. If $w_{i-1} - v_i$ has a negative coordinate, we do not have to compute $\mu_{i-1}/(a_i + b_i\alpha)$.

The condition that $w_{i-1} - v_i$ has non-negative coordinates is not a sufficient condition for $a_i + b_i\alpha$ to divide μ_{i-1}, but it is nearly so. That is, if $w_{i-1} - v_i$ has non-negative coordinates, then the only prime numbers that can divide the denominators of the coefficients of $\mu_{i-1}/(a_i + b_i\alpha)$ are the prime numbers $p \leq y$ that divide $[\mathcal{O} : \mathbf{Z}[\alpha]]$. From Lemma 3.3 it follows that there are only a few such prime numbers, namely not more than $o(\log n)$ for $n \to \infty$. We can modify the procedure described above by always putting $\mu_i = \mu_{i-1}/(a_i + b_i\alpha)$ when $w_{i-1} - v_i$ has non-negative coordinates. Then we have to keep track of the exponents to which those few prime numbers occur in the denominator of μ_i.

The use of exponent vectors suggests that it may be advantageous to order the set S in such a way that the event that $w_i - v_{i-1}$ has non-negative coordinates is frequent. One possible ordering is the one that puts the smoother elements of S first. There may be better orderings than this, but we are not sure what to suggest.

The practical value of these ideas is unclear; the final verdict must await an implementation.

10. ANALYTIC INTERLUDE

In this section we prove a theorem in analytic number theory that is helpful in the complexity analysis of many factoring algorithms, including the number field sieve.

For $x \geq 1$, $y \geq 1$ let $\psi(x, y)$ denote the number of y-smooth positive integers up to x. Suppose x, y are positive integers and consider a process where we choose random integers with the uniform distribution from $[1, x]$ and stop when we have chosen y not necessarily distinct numbers that are y-smooth. The probability that we choose a y-smooth number on one draw is $\psi(x, y)/x$. Thus the expected number of draws to choose y numbers that are y-smooth is $xy/\psi(x, y)$. We now ask for the value of y that minimizes an expression slightly more general than this expectation. Recall the definition of $L_x[u, v]$ from Section 1.

Theorem 10.1. *Suppose g is a function defined for all $y \geq 2$ that satisfies $g(y) \geq 1$ and $g(y) = y^{1+o(1)}$ for $y \to \infty$. Then as $x \to \infty$,*

$$\frac{xg(y)}{\psi(x, y)} \geq L_x[\tfrac{1}{2}, \sqrt{2} + o(1)]$$

uniformly for all $y \geq 2$. In addition,

$$\frac{xg(y)}{\psi(x, y)} = L_x[\tfrac{1}{2}, \sqrt{2} + o(1)]$$

for $x \to \infty$ if and only if $y = L_x[\tfrac{1}{2}, \sqrt{2}/2 + o(1)]$ for $x \to \infty$.

Proof. We shall use the following result from [8]. For any $\epsilon > 0$ we have

(10.2) $\psi(x, x^{1/w}) = x/w^{(1+o(1))w}$ for $w \to \infty$,

uniformly in the region $x \geq w^{(1+\epsilon)w}$.
    We first show that if $y \leq L_x[\tfrac{1}{2}, \tfrac{1}{4}]$ or $y \geq L_x[\tfrac{1}{2}, 2]$, then

(10.3) $\dfrac{xg(y)}{\psi(x, y)} \geq L_x[\tfrac{1}{2}, 2 + o(1)]$ for $x \to \infty$.

Indeed, if $y \leq L_x[\tfrac{1}{2}, \tfrac{1}{4}]$, then (10.2) implies that

$$\frac{xg(y)}{\psi(x, y)} \geq \frac{x}{\psi(x, y)} \geq \frac{x}{\psi(x, L_x[\tfrac{1}{2}, \tfrac{1}{4}])} = L_x[\tfrac{1}{2}, 2 + o(1)]$$

for $x \to \infty$. If $y \geq L_x[\tfrac{1}{2}, 2]$, then it is clear that (10.3) holds since $x/\psi(x, y) \geq 1$.
    Note that (10.2) implies that if $y = L_x[\tfrac{1}{2}, \vartheta]$, then

(10.4) $\dfrac{xg(y)}{\psi(x, y)} = L_x[\tfrac{1}{2}, \vartheta + 1/(2\vartheta) + o(1)]$ for $x \to \infty$

uniformly for ϑ in any compact subset of the set of positive real numbers. Further $\vartheta + 1/(2\vartheta)$ has its minimum value for $\vartheta > 0$ at $\vartheta = \sqrt{2}/2$ and nowhere else. This minimum value is $\sqrt{2}$, which proves the theorem.
    Theorem 10.1 is useful in the analysis of many factoring algorithms. For example, suppose an algorithm factoring n produces auxiliary numbers up to $x = x(n)$ and hopes to find $y^{1+o(1)}$ (for $n \to \infty$) auxiliary numbers that are y-smooth. If these auxiliary numbers are just as likely to be y-smooth as random integers up to x, then we expect to examine $xy^{1+o(1)}/\psi(x, y)$ of these to find the y-smooth integers that we need. If the time to test a single auxiliary number for y-smoothness is $y^{o(1)}$, the expected time for this stage of the factoring algorithm is $xy^{1+o(1)}/\psi(x, y)$. Theorem 10.1 tells us how to choose y so as to minimize this running time, namely $y = L_x[\tfrac{1}{2}, \sqrt{2}/2 + o(1)]$. Further, this running time would be $y^{2+o(1)} = L_x[\tfrac{1}{2}, \sqrt{2} + o(1)]$. Thus if other steps in the algorithm, such as processing a matrix, also take time at most $y^{2+o(1)}$, then $L_x[\tfrac{1}{2}, \sqrt{2} + o(1)]$ is the running time of the complete algorithm. This leads to the following heuristic principle: *if x is a bound on the numbers that "would be smooth" in a factoring algorithm, then the running time of the algorithm is $L_x[\tfrac{1}{2}, \sqrt{2} + o(1)]$.*

For some factoring algorithms, this outline of a complexity analysis can be used as the backbone of a completely rigorous analysis, such as with Examples 10.5, 10.6 and 10.7 below. For other factoring algorithms, the above argument is supplemented with various heuristic assumptions, one of which is often that the auxiliary numbers that "would be smooth" are just as likely to be smooth as random integers of the same approximate magnitude.

Example 10.5. In the random squares algorithm of Dixon (see [12]) the bound for the auxiliary numbers that would be smooth is $x = n$. The running time of the algorithm thus turns out to be $L_n[\frac{1}{2}, \sqrt{2} + o(1)]$ (see [34]). Here, and in the next two examples, we use the elliptic curve smoothness test (see [28; 34]) so that most y-smooth numbers can (rigorously) be recognized to be y-smooth in time $y^{o(1)}$.

Example 10.6. In [36], Vallée modified the random squares method so that the bound for the auxiliary numbers that would be smooth is $x = n^{2/3 + o(1)}$. Thus the running time for her algorithm is $L_n[\frac{1}{2}, \sqrt{4/3} + o(1)]$.

Example 10.7. In the class group relations method [28] the size of the numbers that would be smooth is $n^{1/2 + o(1)}$, and its running time is $L_n[\frac{1}{2}, 1 + o(1)]$.

Example 10.8. In the quadratic sieve method [33] the size of the numbers that would be smooth is $n^{1/2 + o(1)}$ and so its heuristic running time is $L_n[\frac{1}{2}, 1 + o(1)]$. Here sieving replaces the elliptic curve method as a smoothness test.

The heuristic even works for the elliptic curve factoring method [26]. Here the auxiliary numbers that would be smooth are near the least prime factor p of n. We need to find only one y-smooth auxiliary number, but the time to process one trial is not $y^{o(1)}$ but $y^{1+o(1)}$. Thus the heuristic expected time is still $xy^{1+o(1)}/\psi(x, y)$ where $x = p$. Hence Theorem 10.1 applies and we find that the heuristic running time of the elliptic curve method to factor n is $L_p[\frac{1}{2}, \sqrt{2} + o(1)]$ arithmetic operations with integers the size of n.

A sixth example is provided by the number field sieve. Its heuristic complexity analysis, which is given in Section 11, depends on the two final results of this section.

Lemma 10.9. *For real numbers $k \geq e$, $l \geq 1$, define the number $v = v(k, l)$ by*

$$\frac{v^2}{\log v} = kv + l, \qquad v \geq e.$$

Then we have

$$2v = (1 + o(1))\left(k \log k + \sqrt{(k \log k)^2 + 2l \log l}\right)$$

as $k + l \to \infty$.

Proof. From $v((v/\log v) - k) = l$ one sees that v is well-defined and that $v \to \infty$ as $k + l \to \infty$. To prove the lemma, we shall show that we can transform the defining equation

$$(10.10) \qquad v^2 = kv \log v + l \log v$$

into the quadratic equation

$$(10.11) \qquad v^2 = (1 + o(1))(kv \log k + \frac{l \log l}{2}) \qquad \text{as} \quad k + l \to \infty.$$

We distinguish two cases. First suppose that $kv \geq l$, say $kv = cl$ with $c \geq 1$. Then from $kv \leq v^2/(\log v) \leq 2kv$ it follows that $k \to \infty$ and $\log v = (1 + o(1)) \log k$ as $k + l \to \infty$. Hence the first term on the right of (10.10) is $(1 + o(1))kv \log k$. Using that $l = kv/c$ and that $\log c = O(c)$, we see that the second term is

$$l \log v = \frac{l \log l}{2} + \frac{kv}{2c}(\log v - \log k + \log c) = \frac{l \log l}{2} + o(kv \log k).$$

This gives (10.11). In the second case we have $l \geq kv$, say $l = ckv$ with $c \geq 1$. Then from $l \leq v^2/\log v \leq 2l$ we obtain $\log v = (\frac{1}{2} + o(1)) \log l$. The second term on the right of (10.10) is then $(1 + o(1))(l \log l)/2$, and the first is

$$kv \log v = kv \log k + \frac{l}{c}(2 \log v - \log l + \log c) = kv \log k + o(l \log l).$$

This gives again (10.11). Solving the quadratic equation we obtain the lemma.

Lemma 10.12. *For each pair of positive integers n, d satisfying $n > d^{2d^2} > 1$, let real numbers $u = u(n, d) \geq 2$ and $y = y(n, d) \geq 2$ be given, with the property that the number*

$$x = x(n, d) = 2dn^{2/d}u^{d+1}$$

satisfies

$$(10.13) \qquad \frac{u^2 \psi(x, y)}{x} \geq g(y)$$

for some function g satisfying $g(y) \geq 1$ and $g(y) = y^{1+o(1)}$ as $y \to \infty$. Then we have

$$2 \log u \geq (1 + o(1))\left(d \log d + \sqrt{(d \log d)^2 + 4 \log(n^{1/d}) \log \log(n^{1/d})}\right)$$

for $n \to \infty$, uniformly in d.

Proof. In the proof, all $o(1)$'s are for $n \to \infty$, uniformly in d. From $x^x > n$ we see that $x \to \infty$ as $n \to \infty$. Hence Theorem 10.1 implies that

$$u^2 \geq \frac{xg(y)}{\psi(x, y)} \geq L_x[\tfrac{1}{2}, \sqrt{2} + o(1)].$$

Taking the square of the logarithm on both sides we obtain

$$2(\log u)^2 \geq (1 + o(1)) \log x \log \log x.$$

Dividing each side by its logarithm, and using that $t/\log t$ is an increasing function of t for $t \geq e$, we find that

$$\frac{(\log u)^2}{\log \log u} \geq (1 + o(1)) \log x = (1 + o(1))(\frac{2}{d} \log n + (d + 1) \log u).$$

Applying 10.9, with $k \geq (1 + o(1))(d + 1)$ and $l \geq (2 + o(1)) \log(n^{1/d})$, we obtain the lemma.

11. SUMMARY OF THE NUMBER FIELD SIEVE AND A HEURISTIC ANALYSIS

We are finally in a position to list the steps of the number field sieve with some precision and to analyze its running time.

Algorithm 11.1. Given a positive integer n, together with parameters d, u, and y satisfying $d > 1$ and $n > d^{2d^2}$, this algorithm attempts to find a non-trivial factor of n or to prove that n is prime; it halts whether or not it is successful.

Step 1. Test whether n is a power of a prime (see [23, Section 2]) or is divisible by a prime that is less than or equal to y. In either case, output the prime and stop.

Step 2. Apply the base m algorithm (see Section 3) to find an integer m and a monic polynomial $f \in \mathbf{Z}[X]$ of degree d such that $f(m) \equiv 0 \bmod n$. Factor f into irreducible factors in $\mathbf{Z}[X]$ by the algorithm of [22]. If f is found to be reducible, with non-trivial factor g, output the non-trivial factor $g(m)$ of n and stop. Assume now that f is irreducible, and denote by α a zero of f. Compute $\gcd(f'(m), n)$. If this is a non-trivial factor of n output this factor and then stop.

Step 3. As described in Sections 4 and 5, use a sieve to find all members of the set
$$T = \{(a, b) \in \mathbf{Z}^2 : \gcd(a, b) = 1, \ |a| \leq u,$$
$$0 < b \leq u, \ (a + bm)N(a + b\alpha) \text{ is } y\text{-smooth}\}.$$

Step 4. Form the matrix whose rows are the $\mathbf{F}_2$-vectors $e(a, b)$, as defined in Section 8, for $(a, b) \in T$. Use the Wiedemann coordinate recurrence algorithm (see [41]) to find a non-trivial linear dependence relation on the rows of the matrix. If this is unsuccessful, stop. If it is successful, let S be the set of pairs (a, b) for which $e(a, b)$ occurs in the dependence relation.

Step 5. Express the algebraic integer $\gamma = f'(\alpha)^2 \prod_{(a,b) \in S} (a + b\alpha)$ as a polynomial in α of degree less than d. Attempt to find a square root $\beta = \sum_{i=0}^{d-1} b_i \alpha^i$ of γ by the method of [39] (see Section 9). If this is unsuccessful, stop.

Step 6. For c an integer with $c^2 = f'(m)^2 \prod_{(a,b) \in S} (a + bm)$, find the residue $c \bmod n$.

Step 7. Compute $\gcd(c - \sum_{i=0}^{d-1} b_i m^i, n)$. If this is a non-trivial factor of n, output the result and stop. Otherwise, remove an element of S from T and start again at Step 4.

This completes the description of the algorithm.

The following conjectural result describes the optimal choice of the parameters d, u, and y, and the running time of the algorithm for this choice.

Conjecture 11.2. *For each integer n with $n > 256$, one can choose d, u, and y, such that*
$$d = (3^{1/3} + o(1))(\log n / \log \log n)^{1/3}, \qquad n > d^{2d^2} > 1,$$
$$u = y = L_n[\tfrac{1}{3}, (\tfrac{8}{9})^{1/3} + o(1)]$$

for $n \to \infty$, and such that Algorithm 11.1, on input n, d, u, and y, succeeds either in finding a non-trivial factor of n or in proving that n is prime, in time at most

$$(11.3) \qquad L_n[\tfrac{1}{3}, (64/9)^{1/3} + o(1)]$$

for $n \to \infty$. Moreover, this is optimal in the sense that for general n and for all choices of d, u and y satisfying $n > d^{2d^2} > 1$ for which the algorithm is successful, the expression (11.3) is a lower bound for the time taken by the algorithm.

The adjective "general" in the last assertion of the conjecture is meant to express that we allow for exceptional integers n, for which the algorithm takes less time. For example, if n is a power of a prime number, then Algorithm 11.1 terminates in Step 1 in time much less than (11.3), independently of the choice of d, u, and y. Likewise, if n has a relatively small prime factor, then there may be a choice of y for which the algorithm terminates in Step 1 in time less than (11.3). Next, there is a very small class of integers that for a suitable choice of d are factored in Step 2 with very little effort. Finally, if the coefficients of the polynomial f constructed in Step 2 are, for a suitable value of d, much smaller than their upper bound $n^{1/d}$, then it is reasonable to suppose that one can factor n in time less than (11.3), with values for u and y that may not be those in the conjecture. This occurs, for example, if the special number field sieve [24] can be applied. We do not know whether further categories of exceptional integers n exist, but we believe that most integers divisible by at least two distinct primes and not divisible by any small primes are in the class of "general" integers for which (11.3) is a lower bound for the time taken by Algorithm 11.1 to factor them.

The following more general conjecture describes the optimal choice of u and y for given n and d.

Conjecture 11.4. For any two positive integers n and d satisfying $n > d^{2d^2} > 1$, one can choose u and y such that each of u and y is

$$(11.5) \qquad \exp\left(\left(\tfrac{1}{2} + o(1)\right)\left(d\log d + \sqrt{(d\log d)^2 + 4\log(n^{1/d})\log\log(n^{1/d})}\right)\right)$$

and such that Algorithm 11.1, on input n, d, u, and y, succeeds either in finding a non-trivial factor of n or in proving that n is prime, in time at most

$$(11.6) \qquad \exp\left((1 + o(1))\left(d\log d + \sqrt{(d\log d)^2 + 4\log(n^{1/d})\log\log(n^{1/d})}\right)\right),$$

where the $o(1)$'s are for $n \to \infty$, uniformly in d. Moreover, this is optimal in the sense that for general n, for all d in the region $n > d^{2d^2} > 1$, and for all choices of u, y for which the algorithm is successful, the time taken by the algorithm is at least (11.6).

To deduce 11.2 from 11.4 it suffices to choose d so as to minimize (11.6). It is easy to see that we have to make $(d\log d)^2$ and $\log(n^{1/d})\log\log(n^{1/d})$ of the same order of magnitude, which occurs when d has the same order of magnitude as $(\log n / \log \log n)^{1/3}$. Putting $d = \delta(\log n / \log \log n)^{1/3}$ and optimizing δ we find that the optimal choice of d satisfies $\delta = 3^{1/3} + o(1)$ for $n \to \infty$. This immediately leads to 11.2.

We now present a heuristic argument for the correctness of Conjecture 11.4. We begin with the last assertion of the conjecture, which states that (11.6) is, in general, a lower bound for the running time. We deduce this from Lemma 10.12. If we assume that the algorithm does not terminate in Step 1 or in Step 2, then the running time is at least the total number of locations in the sieve from Section 4 that is used in Step 3, which is at least u^2. The lower bound for u^2 that is given by Lemma 10.12 thus leads immediately to the lower bound (11.6) for the running time, provided that we check that condition (10.13) is satisfied. We shall deduce (10.13), heuristically, from a constraint that is implicit in Step 4 of the algorithm, namely, the condition that the number of rows of the matrix in this step is at least of the same order of magnitude as the number of columns; otherwise Step 4 is unlikely to be successful in finding a set S. The number of columns is at least the number of primes in the factor base on the rational side. This is $\pi(y)$, which is $y^{1+o(1)}$ for $y \to \infty$, as required for the right side of (10.13). To estimate the number of rows, we first discuss a bound on the magnitude of the auxiliary numbers generated in Step 3 that "would be smooth". For $|a| \leq u$ and $0 < b \leq u$, the integer $(a + bm)N(a + b\alpha)$ has absolute value at most

$$(u + um)(d + 1)mu^d \leq 2dm^2 u^{d+1} \leq 2dn^{2/d}u^{d+1},$$

since the coefficients of f are bounded by m and $m \leq n^{1/d}$. Hence the number $x = 2dn^{2/d}u^{d+1}$ defined in Lemma 10.12 is a bound on the auxiliary numbers that would be smooth. A random positive integer up to x is y-smooth with probability $\psi(x, y)/[x]$. The number of integers that we try is the number of pairs of integers a, b satisfying $|a| \leq u$, $0 < b \leq u$, and $\gcd(a, b) = 1$, which is about cu^2 for $c = 12/\pi^2$. Thus we might naively think that a good approximation to the cardinality of the set T in Step 3 is given by $cu^2\psi(x, y)/x$. This belief then leads to (10.13), the constant c being absorbed in the factor $y^{o(1)}$ that we allow on the right hand side of (10.13).

We do not know to what extent the naive belief on which the above argument relies is justified. However, we feel that it is reasonable to suppose that for "general" n, d our approximation to the cardinality of T is correct within an exponent $1 + o(1)$ for $y \to \infty$ (as allowed by (10.13)), at least for the values of u and y that are relevant for the algorithm.

Next we turn to the first assertion of 11.4. Our heuristic argument for this is based on the same naive belief as above. Inspecting the case in which equality is achieved in Lemma 10.12, we find in a straightforward way that the numbers

$$u_0 = y_0 = \exp\left(\tfrac{1}{2}\left(d \log d + \sqrt{(d \log d)^2 + 4\log(n^{1/d})\log\log(n^{1/d})}\right)\right),$$

$$x_0 = 2dn^{2/d}u_0^{d+1}$$

satisfy

$$(11.7) \qquad \frac{u_0^2 \psi(x_0, y_0)}{x_0} = y_0^{1+o(1)},$$

the $o(1)$'s here and in the rest of the argument being for $n \to \infty$, uniformly in d. We shall choose u and y a little larger. Specifically, let ϵ be a positive real number, and put

$$u = y = \exp\left(\frac{1+\epsilon}{2}\left(d\log d + \sqrt{(d\log d)^2 + 4\log(n^{1/d})\log\log(n^{1/d})}\right)\right),$$

$$x = 2dn^{2/d}u^{d+1}.$$

Note that these numbers tend to infinity with n, and that we have $\log n = y^{o(1)}$ and $(\log y)/\log x = o(1)$. From $y = y_0^{1+\epsilon}$, $x \leq x_0^{1+\epsilon}$ we see that $(\log x)/\log y \leq (\log x_0)/\log y_0$, so (10.2) gives

$$\frac{\psi(x,y)}{x} \geq \left(\frac{\psi(x_0, y_0)}{x_0}\right)^{1+o(1)}.$$

Combining this with (11.7) we obtain

$$\frac{u^2\psi(x,y)}{x} \geq \left(\frac{u_0^{2(1+\epsilon)}\psi(x_0,y_0)}{x_0}\right)^{1+o(1)} = \left(u_0^{2\epsilon}y_0\right)^{1+o(1)},$$

which by $u_0 = y_0 = y^{1/(1+\epsilon)}$ implies that

(11.8)
$$\frac{u^2\psi(x,y)}{x} \geq y^{(1+o(1))(1+2\epsilon)/(1+\epsilon)}.$$

From this inequality we shall deduce, heuristically, that there is a constant $n(\epsilon)$ such that for $n > n(\epsilon)$, with the above choices of u and y, the number of rows in the matrix in Step 4 is at least the number of columns in the matrix plus an upper bound for the number of times that we cycle through Steps 4 to 7.

As above, we estimate the number of rows to be $(u^2\psi(x,y)/x)^{1+o(1)}$. The number of columns is, in the notation of Section 8, equal to $1 + B + B' + B''$. We have

$$B = \pi(y) < y, \qquad B' \leq dy, \qquad B'' \leq 5\log n, \qquad d < \log n = y^{o(1)},$$

and therefore

$$1 + B + B' + B'' = y^{1+o(1)}.$$

Finally, the number of times that we cycle through Steps 4 to 7 is one more than the number of times that we find a trivial factor of n in Step 7, which is heuristically bounded by $(\log n)^{O(1)} = y^{o(1)}$. Thus our assertion follows, heuristically, from (11.8).

We conclude that every time that Step 4 is performed, it finds a non-trivial linear relation between the rows of the matrix. The linear relations found by the algorithm are linearly independent, so it is reasonable to conjecture that ultimately one of these relations will give rise to a non-trivial factor of n in Step 7. Letting ϵ tend to 0 for $n \to \infty$ we find that we can indeed choose u and

y such that each of them is given by (11.5) and such that the algorithm is likely
to be successful on input n, d, u, y. Then we have $u = u_0^{1+o(1)}$, $y = y_0^{1+o(1)}$,
so (11.7) is also true with u_0, y_0, x_0 replaced by u, y, x.

It remains to estimate the running time of the algorithm with this choice of
parameters. It is easy to see that the time taken by Step 3 equals $u^{2+o(1)}$, which
is the length of the sieve multiplied by a lower order factor; this gives rise to the
expression (11.6). It is clear that Steps 1, 2, 6 and 7 are negligible compared with
Step 3. To estimate the running time of the Wiedemann coordinate recurrence
method in Step 4, we note that the matrix formed in this step has $y^{1+o(1)}$ columns
and about as many rows. In addition, the number of non-zero entries in each
row is $O(\log n) = y^{o(1)}$. Thus the number of non-zero entries in the matrix is
$y^{1+o(1)}$ and the running time of Step 4 is $y^{2+o(1)}$. This is the same as our bound
for the running time of Step 3. In Section 9 we saw that the running time for
Step 5 is $y^{2+o(1)}$ if we use naive arithmetic and $y^{1+o(1)}$ if we use fast arithmetic
subroutines. Thus either way this step too is dominated by Step 3. Finally, as
we saw above, the number of times that we cycle through Step 4 to 7 is likely
to be $y^{o(1)}$.

This concludes our heuristic argument supporting Conjectures 11.2 and 11.4.
We note that the bound for the numbers that "would be smooth" is

$$x = \exp\left(\left(\tfrac{1}{2}+o(1)\right)\left(d^2\log d + 4\log(n^{1/d}) + d\sqrt{(d\log d)^2 + 4\log(n^{1/d})\log\log(n^{1/d})}\right)\right)$$

for $n \to \infty$, uniformly in d, when u is chosen as in Conjecture 11.4, and

$$x = L_n[\tfrac{2}{3}, (64/3)^{1/3} + o(1)] \qquad \text{for} \quad n \to \infty$$

when d and u are chosen as in Conjecture 11.2.

We make a final remark concerning the numbers $(a+bm)N(a+b\alpha)$ in Step 3
that are examined for y-smoothness. We have assumed above that these integers
are just as likely to be y-smooth as random integers of the same magnitude. In
fact, the alert reader may have noticed that these numbers, since they already
factor into the two smaller numbers $a + bm$ and $N(a + b\alpha)$, perhaps have a
greater chance of being y-smooth than a random integer. For practical purposes
this may be true. Asymptotically, however, an argument similar to the one above,
but taking this factorization into account, can be worked out, and it gives exactly
the same results. That is, any differences in the two analyses are absorbed in the
expression "$o(1)$". It may be of interest to point out that when $(a+bm)N(a+b\alpha)$
is y-smooth and a, b are coprime, then the numbers $a + bm$ and $N(a + b\alpha)$ are
coprime too. Indeed, if a prime p divides both, it divides $f(m) = n$. However,
after Step 1 we are assured that n has only prime factors greater than y.

12. HOMOGENEOUS POLYNOMIALS

In this section we discuss a modification of the number field sieve, in which
the one-variable polynomial f is replaced by a homogeneous polynomial in two
variables. This has the advantage that its coefficients can be taken a bit smaller,
which improves the practical performance of the algorithm. We first describe the
algorithm, and then provide additional explanations for some of the individual
steps.

Algorithm 12.1. Given an integer $n > 1$, which is not a power of a prime number (see 11.1, Step 1), together with parameters d, u, and y, which are positive integers, this algorithm attempts to find a non-trivial factor of n.

Step 1. Find integers m_1, m_2, and a dth degree homogeneous polynomial

$$f = c_d X^d + c_{d-1} X^{d-1} Y + \cdots + c_1 X Y^{d-1} + c_0 Y^d \in \mathbf{Z}[X, Y]$$

such that m_1, m_2, and the coefficients c_i are "small" and such that we have $f(m_1, m_2) \equiv 0 \bmod n$, $f(m_1, m_2) \neq 0$. See 12.2 for methods to select f, m_1, m_2.

Step 2. Check that f is irreducible in $\mathbf{Z}[X, Y]$, so that in particular its content equals 1, and that $f \neq X$, $f \neq Y$. Further check that each of m_2, c_d, and

$$f_X(m_1, m_2) = \sum_{i=1}^{d} i c_i m_1^{i-1} m_2^{d-i}, \qquad \text{where } f_X = \frac{\partial f}{\partial X},$$

is coprime to n. See 12.5 for more information on this step.

Step 3. For each prime number $p \leq y$, determine the set $R'(p)$ of elements $(r_1 : r_2)$ of the projective line $\mathbf{P}^1(\mathbf{F}_p)$ over $\mathbf{F}_p$ for which $f(r_1, r_2) = 0$. Note that if we identify $\mathbf{P}^1(\mathbf{F}_p)$ with $\mathbf{F}_p \cup \{\infty\}$ by identifying $(r_1 : r_2)$ with r_1/r_2, then $R'(p)$ consists of those $r = r_1/r_2 \in \mathbf{F}_p$ for which $f(r, 1) = 0$, together with ∞ if $c_d \equiv 0 \bmod p$.

Step 4. Find all members of the set

$$T = \{(a, b) \in \mathbf{Z}^2 : \gcd(a, b) = 1, |a| \leq u,$$
$$0 < b \leq u, (am_2 - bm_1)f(a, b) \text{ is } y\text{-smooth}\}.$$

This is done with a sieve, as described in Sections 4 and 5. Note that, for coprime integers a, b, and a prime number p, the number $f(a, b)$ is divisible by p if and only if $(a \bmod p : b \bmod p) \in R'(p)$.

Step 5. For each $(a, b) \in T$, form the $\mathbf{F}_2$-vector $e(a, b)$ that is defined as follows. The first coordinate of $e(a, b)$ is determined by the sign of $am_2 - bm_1$, as in Section 8 (we cannot have $am_1 - bm_2 = 0$ if $\max\{|m_1|, |m_2|\} > u$, which will be the case with our choice of parameters). The next $B = \pi(y)$ coordinates are given by $\text{ord}_p(am_2 - bm_1) \bmod 2$, as p runs over the prime numbers $\leq y$. Next there are B' coordinates, where $B' = \sum_{p \leq y} \#R'(p)$, the sum ranging over prime numbers p. For each prime number $p \leq y$ and each $r \in R'(p)$, the (p, r)th coordinate of $e(a, b)$ is equal to $e_{p,r}(a, b) \bmod 2$, where $e_{p,r}(a, b)$ equals $\text{ord}_p(f(a, b))$ if $(a \bmod p : b \bmod p) = r$ and $e_{p,r}(a, b) = 0$ otherwise. Each of the following B'' coordinates corresponds to a prime number $q > y$ and a pair of numbers s_1, s_2 with $(s_1 \bmod q : s_2 \bmod q) \in R'(q)$; see 12.7 for the choice of B'' and the triples q, s_1, s_2. The (q, s_1, s_2)th coordinate of $e(a, b)$ is 0 mod 2 if the Legendre symbol $\left(\frac{as_2 - bs_1}{q}\right)$ equals 1, and 1 mod 2 if it equals -1. Finally, $e(a, b)$ has a last coordinate that is equal to 1 mod 2. (Dan Bernstein points out that this last coordinate can be omitted if $m_2 = 1$ and $m_1 > u$, since then it is equal to the first coordinate, which gives the sign of $am_2 - bm_1$.) Thus $e(a, b) \in \mathbf{F}_2^{2+B+B'+B''}$.

Step 6. Use the Wiedemann coordinate recurrence algorithm (see [41]) to find a non-trivial linear dependence relation between the vectors $e(a,b)$, $(a,b) \in T$. If this is unsuccessful, stop. If it is successful, let S be the set of pairs (a,b) for which $e(a,b)$ occurs in the dependence relation. Note that $\#S$ is even, due to the presence of the last coordinate.

Step 7. Let the algebraic integer ω be a zero of the polynomial $f(X, c_d)$. Express the algebraic integer

$$\gamma = (f_X(\omega, c_d)/c_d)^2 \cdot \prod_{(a,b) \in S} (c_d a - b\omega)$$

(with f_X as in Step 2) as a polynomial in ω of degree less than d. Attempt to find a square root β of γ by the method of [39] (see Section 9). If this is unsuccessful, stop. Otherwise, if $\beta = \sum_{i=0}^{d-1} b_i \omega^i$, with $b_i \in \mathbf{Z}$, calculate an integer v with $v \equiv \sum_{i=0}^{d-1} b_i c_d^i m_1^i m_2^{d-1-i} \bmod n$.

Step 8. For w an integer with $w^2 = \prod_{(a,b) \in S} (am_2 - bm_1)$, find the residue $w \bmod n$. In addition, calculate integers h, l with

$$h \equiv c_d^{d-2+\#S/2} \cdot f_X(m_1, m_2) \bmod n, \qquad l \equiv m_2^{\#S/2} \bmod n.$$

Step 9. Compute $\gcd(hw - lv, n)$. If this is a non-trivial factor of n, output the result and stop. Otherwise, remove an element of S from T and start again at Step 6.

This completes the description of the algorithm. We now discuss some of the individual steps.

12.2. *Selecting f and m_1, m_2.* If we insist on choices for which $m_2 = c_d = 1$, then Algorithm 12.1 reduces to 11.1, except for the last coordinate that was appended to the vectors $e(a,b)$. We now discuss three methods for choosing f, m_1, m_2. In the first method we allow $c_d \neq 1$, in the second method we allow $m_2 \neq 1$, and in the third method we allow both.

In the first method we take $m_2 = 1$, and we let m_1 be the least integer exceeding $n^{1/(d+1)}$. We obtain f by expanding n in base m_1, so that

$$n = c_d m_1^d + c_{d-1} m_1^{d-1} + \cdots + c_1 m_1 + c_0, \qquad 0 \le c_i < m_1.$$

With this method, we have $|m_i| \le n^{1/(d+1)} + 1$, $|c_i| \le n^{1/(d+1)}$. Of course, we can modify this method by changing m_1 a little, by allowing some of the digits c_i to be negative, or by replacing n by a small multiple.

In the second method we take $c_d = 1$. To find the other c_i and m_1, m_2, we proceed as follows. For m_1 one tries several values with $m_1 \approx n^{1/(d+1)}$, until one discovers a value for which $n - m_1^d$ is found to have a divisor m_2 with $m_2 \approx n^{1/(d+1)}$; for example, by trial division or by the elliptic curve method one may discover so many small factors of $n - m_1^d$ that it is easy to multiply some of them together in order to obtain a suitable m_2. Note that $\gcd(m_1, m_2)$ divides n

and is generally much smaller than n; so we may assume that $\gcd(m_1, m_2) = 1$. Next one determines small coefficients c_i such that

$$(12.3) \qquad \frac{n - m_1^d}{m_2} = c_{d-1}m_1^{d-1} + \cdots + c_1 m_1 m_2^{d-2} + c_0 m_2^{d-1}.$$

One can do this either by going from the right, determining $c_0, c_1, \ldots$ successively by looking modulo m_1 and requiring that $|c_i| \le m_1/2$ (or $0 \le c_i < m_1$); or similarly from the left and finding $c_{d-1}, c_{d-2}, \ldots$ from congruences modulo m_2; or by determining some from the right and some from the left. In all cases, the final c_j to be determined is forced by equation (12.3). If d is small in comparison with $n^{1/(d+1)}$, as it will be in practice, then the order of magnitude of $|c_j|$ will not be much larger than $n^{1/(d+1)}$. Again, this method allows several refinements. For example, one might choose m_1, m_2 to be $\approx (n/d)^{1/(d+1)}$; a judicious choice of non-negative values for the c_i may then result in a smaller value for the final c_j.

In the third method we allow both $m_2 \ne 1$ and $c_d \ne 1$. Although we do not know how to exploit this freedom in order to obtain substantially better results, it is still of interest to see how one can proceed. Namely, one can first choose arbitrary coprime integers m_1, m_2 that are $\approx n^{1/(d+1)}$. Next one needs to determine the c_i such that

$$(12.4) \qquad kn = c_d m_1^d + c_{d-1} m_1^{d-1} m_2 + \cdots + c_0 m_2^d$$

for some small non-zero integer k. One can either do this by first choosing k (for example, $k = 1$), and next determining the c_i by one of the methods that we indicated for solving (12.3). Alternatively, one can consider the subgroup

$$L = \{(x_i)_{i=0}^d : \sum_{i=0}^d x_i m_1^i m_2^{d-i} \equiv 0 \bmod n\}$$

of $\mathbf{Z}^{d+1}$. A basis of L is given by $(0, 0, \ldots, 0, n)$ together with the d vectors

$$(0, \ldots, 0, 1, -t, 0, \ldots, 0),$$

where $t \in \mathbf{Z}$ is such that $tm_2 \equiv m_1 \bmod n$ (here we assume that $\gcd(m_2, n) = 1$; see 12.5). One can apply a lattice basis reduction algorithm (see [22]) to find a basis of L that consists of relatively short vectors. At least one of the vectors $(x_i)_{i=0}^d$ in the reduced basis satisfies $\sum_{i=0}^d x_i m_1^i m_2^{d-i} \ne 0$, and a solution to (12.4) is then given by $c_i = x_i$. Also for this algorithm one expects the c_i to be of order of magnitude $n^{1/(d+1)}$.

In the above we attempted to minimize the absolute values of m_1, m_2, and the coefficients of f. It should be kept in mind, however, that other properties of f also influence the running time of the algorithm. For example, one may want to choose f in such a way that $f \bmod p$ has many linear factors in $\mathbf{F}_p[X]$ for several small prime numbers p. This increases the smoothness probability of the numbers $f(a, b)$.

12.5. *Irreducibility testing.* With any reasonable choices that are made in Step 1, each of m_2 and c_d will be much less than n in absolute value. Hence if any of $\gcd(m_2, n)$, $\gcd(c_d, n)$ is found to be different from 1 then it is a non-trivial divisor of n, and the algorithm can stop. Assume now that $\gcd(m_2, n) = \gcd(c_d, n) = 1$. The content $\operatorname{cont} f = \gcd(c_0, c_1, \ldots, c_d)$ of the polynomial f divides the multiple $f(m_1, m_2)$ of n, and it is coprime to n because it divides c_d. Therefore the polynomial $f^* = f / \operatorname{cont} f$ still has the property that $f^*(m_1, m_2)$ is divisible by n. Thus, replacing f by f^* if necessary, we may assume that $\operatorname{cont} f = 1$. We can now factor f into irreducible factors in $\mathbf{Z}[X, Y]$ with the algorithm of [22]; note that the factorization of f can easily be obtained from the factorization of the one-variable polynomial $f(X, 1)$. Suppose first that f is found to be reducible, $f = gh$ (say). Then we have $f(m_1, m_2) = g(m_1, m_2)h(m_1, m_2)$, which leads to a splitting of n. For most reasonable choices of the parameters it is very likely that $g(m_1, m_2)$ and $h(m_1, m_2)$ are less than n in absolute value, so that this is a non-trivial splitting. If nevertheless the splitting is trivial, then one of $g(m_1, m_2)$, $h(m_1, m_2)$ is divisible by n, say the first one. Then we can replace f and d by g and $\deg g$. It is easy to see that this replacement improves the algorithm. Let it next be assumed that f is irreducible. Again, in most cases the number $f_X(m_1, m_2)$ will be less than n, so that $\gcd(f_X(m_1, m_2), n)$ is a non-trivial divisor of n if it is not 1; and if it ever happens that $\gcd(f_X(m_1, m_2), n) = n$ then one has the option of replacing f and d by f_X and $d - 1$. Finally, the conditions $f \neq X$, $f \neq Y$ are satisfied if $|m_1|$, $|m_2| < n$, which is very likely to be true.

12.6. *First degree primes.* In Section 5 we saw that the pairs consisting of a prime number p and an element $r \in R(p)$ correspond to the first degree primes of the ring $\mathbf{Z}[\alpha]$. The pairs consisting of a prime number p and an element $r \in R'(p)$ that occur in Algorithm 12.1 can be interpreted in a similar manner. We introduce some notation.

Let $\alpha = \omega / c_d$, where ω is, as in Step 7, a zero of $f(X, c_d)$; so α is a zero of $f(X, 1)$. Note that α is not an algebraic integer unless $c_d = \pm 1$; but ω is an algebraic integer, since $f(X, c_d)/c_d$ is a monic polynomial with integral coefficients. Let the elements $\beta_0, \ldots, \beta_{d-1} \in \mathbf{Z}[\alpha]$ be defined by $f(X, 1)/(X - \alpha) = \sum_{i=0}^{d-1} \beta_i X^i$; so $\beta_i = c_d \alpha^{d-1-i} + c_{d-1}\alpha^{d-2-i} + \cdots + c_{i+1}$. Further let $A = \mathbf{Z} + \sum_{i=0}^{d-2} \mathbf{Z}\beta_i$. A simple computation shows that A is closed under multiplication, so that A is an order in the number field $K = \mathbf{Q}(\alpha)$, in the sense of Section 7 (cf. [27, 2.10]). We have $\mathbf{Z}[\omega] \subset A \subset \mathbf{Z}[\alpha]$, where $\mathbf{Z}[\omega]$ is also an order in K, but $\mathbf{Z}[\alpha]$ is not (unless $c_d = \pm 1$). The discriminant of A is equal to the discriminant Δ of $f(X, 1)$, and the discriminant of $\mathbf{Z}[\omega]$ equals $c_d^{(d-1)(d-2)}\Delta$. It is of interest to observe that the ring A does not change if $f(X, Y)$ is replaced by $f(Y, X)$ and α by α^{-1}; so we also have $A \subset \mathbf{Z}[\alpha^{-1}]$, and in fact one can show that $A = \mathbf{Z}[\alpha] \cap \mathbf{Z}[\alpha^{-1}]$.

With this notation, the pairs consisting of a prime number p and an element $(r_1 : r_2) \in R'(p)$ are in bijective correspondence with the first degree primes $\mathfrak{p}$ of A. If $r_2 \neq 0$, then $\mathfrak{p}$ is the intersection of A and the kernel of the ring homomorphism $\mathbf{Z}[\alpha] \to \mathbf{F}_p$ that sends α to r_1/r_2. If $r_2 = 0$, then $\mathfrak{p}$ is the intersection of A and the kernel of the ring homomorphism $\mathbf{Z}[\alpha^{-1}] \to \mathbf{F}_p$ that sends α^{-1} to 0. Each prime $\mathfrak{p}$ of A gives rise to a function $l_{\mathfrak{p}}$ as in 7.1.

Let a, b be a pair of coprime integers. Then the following analogue of 5.5 is valid. First, if $\mathfrak{p}$ is a prime of A of degree greater than 1, then $l_{\mathfrak{p}}(a - b\alpha) = 0$. Next, let $\mathfrak{p}$ be a first degree prime of A, corresponding to a pair p, $r \in R'(p)$. Then the number $e_{p,r}(a, b)$ that is defined as in Step 5 is equal to the number of composition factors of the A-module $(A + A\alpha)/A(a - b\alpha)$ that are isomorphic to $A/\mathfrak{p}$; explicitly speaking, one has

$$e_{p,r}(a, b) = \begin{cases} l_{\mathfrak{p}}(a - b\alpha) & \text{if } r \neq \infty, \\ l_{\mathfrak{p}}(a - b\alpha) + \operatorname{ord}_p c_d & \text{if } r = \infty. \end{cases}$$

(Note that this is consistent with 7.1(c), since $f(a, b) = c_d N(a - b\alpha)$.) It follows that the analogue of 5.3 holds, provided that we restrict attention to sets S for which $\#S$ is even.

12.7. *Making squares.* It is the purpose of Steps 5 and 6 to find a non-empty subset $S \subset T$ of even cardinality such that $\prod_{(a,b)\in S}(am_2 - bm_1)$ is a square in $\mathbf{Z}$ and $\prod_{(a,b)\in S}(a - b\alpha)$ is a square in K. Clearly, the condition that S be even is taken care of by the last coordinate of the vectors $e(a, b)$, and the condition that the product of the elements $am_2 - bm_1$ be a square by the first $1 + B$ coordinates. The B' coordinates that correspond to the pairs p, r guarantee, by 12.6, that the set S found in Step 6 is such that the product of the elements $a - b\alpha$, for $(a, b) \in S$, belongs to the subgroup V_A of K^* defined in Section 7. One has $V_A \supset K^{*2}$, and depending on the algorithm used for Step 1 one can mimic the proof of Theorem 6.7 and find a constant c for which the obstruction group V_A/K^{*2} has $\mathbf{F}_2$-dimension at most $c \log n$. To overcome this obstruction group, one can use quadratic characters for, say, $B'' = [(c + 2/\log 2)\log n]$ first degree primes $\mathfrak{q}$ of A. As in Section 8, one can choose these primes to be the first B'' primes of norm exceeding y that do not contain $f_X(\omega, c_d)$. Explicitly, one can take the first B'' triples q, s_1, $s_2 = 1$ for which q is a prime number not dividing c_d with $q > y$, and $s_1 \pmod{q}$ is such that $f(s_1, 1) \equiv 0 \bmod q$, $f_X(s_1, 1) \not\equiv 0 \bmod q$, and use these in Step 5.

12.8. *The final congruence.* Suppose that $\prod_{(a,b)\in S}(a - b\alpha)$ is a square in K and that $\#S$ is even. Multiplying by $c_d^{\#S}$ we see that the element $\prod_{(a,b)\in S}(c_d a - b\omega)$ of the order $\mathbf{Z}[\omega]$ is also a square in K. The square root is in the ring of integers of K, so $f_X(\omega, c_d)/c_d$ times the square root belongs to $\mathbf{Z}[\omega]$ (see [40, Proposition 3-7-14]). Hence the element β calculated in Step 7 has coefficients b_i in $\mathbf{Z}$.

Let now the ring homomorphism $\varphi: \mathbf{Z}[\alpha] \to \mathbf{Z}/n\mathbf{Z}$ be such that $\varphi(\alpha) = (m_1 \bmod n)/(m_2 \bmod n)$. Then $\varphi(m_2\omega) = (c_d m_1 \bmod n)$, so with v as in Step 7 we have

$$\varphi(m_2^{d-1}\beta) = (v \bmod n).$$

With l, w and h as in Step 8, this leads to

$$
\begin{aligned}
(l^2 v^2 \bmod n) &= \varphi(m_2^{2(d-1)+\#S} \beta^2) \\
&= \varphi\left(\left(m_2^{d-1} f_X(\omega, c_d)/c_d \right)^2 \cdot \prod_{(a,b) \in S} m_2(c_d a - b\omega) \right) \\
&= \varphi\left(\left(f_X(m_2\omega, m_2 c_d)/c_d \right)^2 \cdot \prod_{(a,b) \in S} c_d(am_2 - bm_1) \right) \\
&= \varphi((c_d^{d-2} f_X(m_1, m_2))^2 c_d^{\#S} w^2) \\
&= (h^2 w^2 \bmod n).
\end{aligned}
$$

This explains the attempt in Step 9 to find a non-trivial factor of n.

12.9. Choice of parameters u, y, d. The heuristic analysis of Algorithm 11.1 given in Section 11 can be copied without essential changes for Algorithm 12.1. The main difference is that the factor $n^{2/d}$ in x is to be replaced by $n^{2/(d+1)}$. Since our analysis gave the optimal value for d only up to a factor $1 + o(1)$ (for $n \to \infty$), the heuristic asymptotic results for Algorithm 12.1 are the same as for Algorithm 11.1. From a practical point of view, 12.1 may be better than 11.1; see the discussion in 12.15.

12.10. The optimal choice of the polynomial. In Section 3 and in 12.2 we described altogether four methods for selecting f, m_1, and m_2. One may ask whether there is a better method for doing this. We present an argument that leads to a limit on the performance of any method for selecting f, m_1, and m_2. It shows that asymptotically one cannot expect to do better than the methods that we described if one wishes the algorithm to apply to all integers n. In addition, the argument suggests that for practical purposes there is still room for improvement (see 12.15).

For a given choice of n and d, what would be a good choice of f, m_1, m_2 in Step 1 of Algorithm 12.1? Let $M = \max\{|m_1|, |m_2|\}$ and let $C = \max\{|c_0|, |c_1|, \ldots, |c_d|\}$. An upper bound on the integers $|(am_2 - bm_1)f(a, b)|$ that are examined for smoothness in Step 4 of the algorithm is $2(d+1)u^{d+1}CM$. Thus for a given n and d, a choice of f, m_1, m_2 which has the product CM small should be better for factoring n than another choice with CM large.

For example, in the base m method used in Algorithm 11.1 we have $M \leq n^{1/d}$ and $C \leq n^{1/d}$, so that $CM \leq n^{2/d}$. The methods of 12.2 achieve $CM = O(n^{2/(d+1)})$, so we would expect these methods to give an improvement over the base m method. The following result expresses that we cannot expect to get CM substantially smaller than $n^{2/(d+2)}$ for all n.

Given positive integers d, C, M, let $\mathcal{S}(d, C, M)$ denote the set of non-zero integers of the form $f(m_1, m_2)$ where m_1, m_2 are integers with $|m_1|, |m_2| \leq M$ and $f = \sum_{i=0}^{d} c_i X^i Y^{d-i} \in \mathbf{Z}[X, Y]$ satisfies $|c_i| \leq C$ for $0 \leq i \leq d$.

Proposition 12.11. *For each $\epsilon > 0$ there is a number $N(\epsilon)$ with the following property. Suppose d, C, M, N are positive integers with $N > N(\epsilon)$. If each* *integer in the interval $[1, N]$ has a multiple in $\mathcal{S}(d, C, M)$, then*

$$CM \geq \frac{1}{8} N^{(2-\epsilon)/(d+2)}.$$

Proof. It suffices to prove the proposition in the case $0 < \epsilon < 1$. Suppose d, C, M, N are positive integers and $\mathcal{S}(d, C, M)$ contains a multiple of each of the integers in $[1, N]$. We may assume that $CM \leq N^{2/d}$ for otherwise there is nothing to prove. It is clear that each member of $\mathcal{S}(d, C, M)$ has absolute value at most $(d+1)CM^d$. Thus

$$(12.12) \qquad N \leq (d+1)CM^d \leq (d+1)(CM)^d \leq (d+1)N^2.$$

Let $D = \max\{\tau(j) : 1 \leq j \leq (d+1)N^2\}$, where $\tau(j)$ denotes the number of divisors of j. Since $\tau(j) = j^{o(1)}$ for $j \to \infty$, there is some number $N(\epsilon)$ such that if $N > N(\epsilon)$ we have

$$(12.13) \qquad D \leq (d+1)^\epsilon N^\epsilon.$$

By our assumption on N we have

$$(12.14) \quad N \leq D \cdot \#\mathcal{S}(d, C, M) \leq D(2C+1)^{d+1}(2M+1)^2 \leq 3^{d+3} DC^{d+1} M^2.$$

Multiplying this by the first inequality in (12.12) we get

$$N^2 \leq 3^{d+3}(d+1)DC^{d+2}M^{d+2},$$

so that using (12.13) we obtain

$$\begin{aligned}
CM &\geq (3^{d+3}(d+1)D)^{-1/(d+2)} N^{2/(d+2)} \\
&\geq 3^{-(d+3)/(d+2)}(d+1)^{-(1+\epsilon)/(d+2)} N^{(2-\epsilon)/(d+2)} \\
&\geq 3^{-4/3} 4^{-2/5} N^{(2-\epsilon)/(d+2)} > \frac{1}{8} N^{(2-\epsilon)/(d+2)}.
\end{aligned}$$

This completes the proof of the proposition.

If we do not require that every integer up to N have a multiple in $\mathcal{S}(d, C, M)$, but only that N does, we still have (12.12) holding, which gives $CM \geq (d+1)^{-1/d}N^{1/d} \geq \frac{1}{2}N^{1/d}$. This lower bound for CM is almost achieved in the *special number field sieve*, which accounts for its lower complexity.

12.15. *Practical considerations.* In practical circumstances, when n is fixed rather than tending to infinity, the above argument suggests that our methods for selecting f, m_1, m_2 are not yet optimal. Namely, suppose that for a given N and d we ignore lower order factors in (12.12) and (12.14) and solve the equations $N = CM^d = C^{d+1}M^2$ for C and M. This suggests we may be able to choose M near N^s and C near N^t where

$$s = \frac{d}{(d-1)(d+2)}, \qquad t = \frac{d-2}{(d-1)(d+2)}.$$

In fact, suppose we choose $M = [N^s]$, $C = [N^t]$, so that $CM \leq N^{s+t} = N^{2/(d+2)}$. It is likely that for most integers in $[1, N]$ there is a choice of f, m_1, m_2 satisfying $|c_i| \leq C$ and $|m_i| \leq M$. Indeed the total number of such triples is $(2C+1)^{d+1}(2M+1)^2$, which by our choice of C, M is somewhat larger than N. Also, the typical order of magnitude of $|f(m_1, m_2)|$ is CM^d, which is about N. But if we have a set of a little over N "pseudorandom" numbers of order of magnitude N, then it is quite likely that most integers in $[1, N]$ have a multiple in the set. Thus if we are interested in a particular number $n \leq N$, either this choice of values for C, M or perhaps slightly larger ones should suffice. Note that this imprecise argument is purely existential, and that it does not suggest a way of constructing f, m_1, m_2.

Suppose that n lies in a realistic range, like $n \approx 10^{130}$, and that we take $d = 5$. Then Algorithm 11.1 uses $s = t = \frac{1}{5}$, and therefore m and the coefficients of f each have about 26 digits. In Algorithm 12.2 we have $s = t = \frac{1}{6}$, so m_1, m_2 and the coefficients of f have about 22 digits, which is a significant improvement. The above argument suggests that the optimal values would be $s = 5/28$ and $t = 3/28$, in which case the m_i would have about 23 digits and the coefficients of f about 14 digits. Thus for practical purposes there may still be room for improvement.

12.16. *Additional improvements.* We mention two variations of the number field sieve that improve its practical performance, while not affecting the asymptotic analysis. The first is the large prime variation, which was used in the factorization of the ninth Fermat number [23]; see also [25]. The second is the lattice sieve idea of Pollard [32].

REFERENCES

1. L.M. Adleman, *Factoring numbers using singular integers*, Proc. 23rd Annual ACM Symp. on Theory of Computing (STOC) (1991), 64–71.
2. E. Bach, *Explicit bounds for primality testing and related problems*, Math. Comp. **55** (1990), 355–380.
3. N. Boston, W. Dabrowski, T. Foguel, P. Gies, D. Jackson, J. Leavitt, D. Ose, *The proportion of fixed-point-free elements in a transitive permutation group*, Comm. in Algebra, to appear.
4. J. Brillhart, M. Filaseta, A. Odlyzko, *On an irreducibility theorem of A. Cohn*, Can. J. Math. **33** (1981), 1055–1059.
5. J. Brillhart, D.H. Lehmer, J.L. Selfridge, B. Tuckerman, S.S. Wagstaff, Jr., *Factorizations of $b^n \pm 1$, $b = 2, 3, 5, 6, 7, 10, 11, 12$ up to high powers*, second edition, Contemporary Mathematics **22**, Amer. Math. Soc., Providence, 1988.
6. J.A. Buchmann, H.W. Lenstra, Jr., *Decomposing primes in number fields*, in preparation.
7. P.J. Cameron, A.M. Cohen, *On the number of fixed point free elements in a permutation group*, Discrete Math. **106/107** (1992), 135–138.
8. E.R. Canfield, P. Erdős, C. Pomerance, *On a problem of Oppenheim concerning "factorisatio numerorum"*, J. Number Theory **17** (1983), 1–28.
9. J.W.S. Cassels, A. Fröhlich (eds), *Algebraic number theory*, Proceedings of an instructional conference, Academic Press, London, 1967.
10. D. Coppersmith, *Modifications to the number field sieve*, J. Cryptology, to appear; IBM Research Report #RC 16264, Yorktown Heights, New York, 1990.
11. J.-M. Couveignes, *Computing a square root for the number field sieve*, this volume, pp. 95–102.

12. J. D. Dixon, *Asymptotically fast factorization of integers*, Math. Comp. **36** (1981), 255–260.

13. W. Fulton, *Intersection theory*, Springer-Verlag, Berlin, 1984.

14. P. X. Gallagher, *The large sieve and probabilistic Galois theory*, in: H. G. Diamond (ed.), *Analytic number theory*, Proc. Symp. Pure Math. **24**, Amer. Math. Soc., Providence, 1973, 91–101.

15. D. Gordon, *Discrete logarithms in GF(p) using the number field sieve*, SIAM J. Discrete Math. **6** (1993), 124–138.

16. B. Huppert, *Endliche Gruppen I*, Springer-Verlag, Berlin, 1967.

17. D. E. Knuth, *The art of computer programming*, volume 2, *Seminumerical algorithms*, second edition, Addison-Wesley, Reading, Mass., 1981.

18. S. Landau, *Factoring polynomials over algebraic number fields*, SIAM J. Comput. **14** (1985), 184–195.

19. S. Lang, *Algebraic number theory*, Addison-Wesley, Reading, Mass., 1970.

20. A. K. Lenstra, *Factorization of polynomials*, in [29], 169–198.

21. A. K. Lenstra, *Factoring polynomials over algebraic number fields*, in: J. A. van Hulzen (ed.), *Computer algebra*, Lecture Notes in Comput. Sci. **162**, Springer-Verlag, Berlin, 1983, 245–254.

22. A. K. Lenstra, H. W. Lenstra, Jr., L. Lovász, *Factoring polynomials with rational coefficients*, Math. Ann. **261** (1982), 515–534.

23. A. K. Lenstra, H. W. Lenstra, Jr., M. S. Manasse, J. M. Pollard, *The factorization of the ninth Fermat number*, Math. Comp. **61** (1993), to appear.

24. A. K. Lenstra, H. W. Lenstra, Jr., M. S. Manasse, J. M. Pollard, *The number field sieve*, this volume, pp. 11–42. Extended abstract: Proc. 22nd Annual ACM Symp. on Theory of Computing (STOC) (1990), 564–572.

25. A. K. Lenstra, M. S. Manasse, *Factoring with two large primes*, Math. Comp., to appear.

26. H. W. Lenstra, Jr., *Factoring integers with elliptic curves*, Ann. of Math. **126** (1987), 649–673.

27. H. W. Lenstra, Jr., *Algorithms in algebraic number theory*, Bull. Amer. Math. Soc. **26** (1992), 211–244.

28. H. W. Lenstra, Jr., C. Pomerance, *A rigorous time bound for factoring integers*, J. Amer. Math. Soc. **5** (1992), 483–516.

29. H. W. Lenstra, Jr., R. Tijdeman (eds), *Computational methods in number theory*, Mathematical Centre Tracts **154/155**, Mathematisch Centrum, Amsterdam, 1982.

30. M. A. Morrison, J. Brillhart, *A method of factoring and the factorization of F_7*, Math. Comp. **29** (1975), 183–205.

31. J. M. Pollard, *Factoring with cubic integers*, this volume, pp. 4–10.

32. J. M. Pollard, *The lattice sieve*, this volume, pp. 43–49.

33. C. Pomerance, *Analysis and comparison of some integer factoring algorithms*, in [29], 89–139.

34. C. Pomerance, *Fast, rigorous factorization and discrete logarithm algorithms*, in: D. S. Johnson, T. Nishizeki, A. Nozaki, H. S. Wilf (eds), *Discrete algorithms and complexity*, Academic Press, Orlando, 1987, 119–143.

35. O. Schirokauer, *On pro-finite groups and on discrete logarithms*, Ph. D. thesis, University of California, Berkeley, May 1992.

36. B. Vallée, *Generation of elements with small modular squares and provably fast integer factoring algorithms*, Math. Comp. **56** (1991), 823–849.

37. B. L. van der Waerden, *Algebra*, seventh edition, Springer-Verlag, Berlin, 1966.

38. P. S. Wang, *Factoring multivariate polynomials over algebraic number fields*, Math. Comp. **30** (1976), 324–336.

39. P. J. Weinberger, L. P. Rothschild, *Factoring polynomials over algebraic number fields*, ACM Trans. Math. Software **2** (1976), 335–350.

40. E. Weiss, *Algebraic number theory*, McGraw-Hill, New York, 1963; reprinted, Chelsea, New York, 1976.

41. D. Wiedemann, *Solving sparse linear equations over finite fields*, IEEE Trans. Inform. Theory **32** (1986), 54–62.

DEPARTMENT OF MATHEMATICS, REED COLLEGE, PORTLAND, OR 97202, U.S.A.
E-mail address: jpb@reed.edu

DEPARTMENT OF MATHEMATICS, UNIVERSITY OF CALIFORNIA, BERKELEY, CA 94720, U.S.A.
E-mail address: hwl@math.berkeley.edu

DEPARTMENT OF MATHEMATICS, UNIVERSITY OF GEORGIA, ATHENS, GA 30602, U.S.A.
E-mail address: carl@ada.math.uga.edu

COMPUTING A SQUARE ROOT
FOR THE NUMBER FIELD SIEVE

JEAN-MARC COUVEIGNES

ABSTRACT. The number field sieve is a method proposed by Lenstra, Lenstra, Manasse and Pollard for integer factorization (this volume, pp. 11–42). A heuristic analysis indicates that this method is asymptotically faster than any other existing one. It has had spectacular successes in factoring numbers of a special form. New technical difficulties arise when the method is adapted for general numbers (this volume, pp. 50–94). Among these is the need for computing the square root of a huge algebraic integer given as a product of hundreds of thousands of small ones. We present a method for computing such a square root that avoids excessively large numbers. It works only if the degree of the number field that is used is odd. The method is based on a careful use of the Chinese remainder theorem.

1. INTRODUCTION

We begin by recalling the basic scheme of the number field sieve, cf. [7]. Let n be a positive integer that is not a power of a prime number. In order to factor n, we first find many congruences modulo n involving a given set of numbers called the *basis*. This is done by means of a suitable ring of algebraic integers. To construct this ring, one chooses a positive integer d, which is the degree of the ring to be constructed; if n has between 110 and 160 decimal digits then $d = 5$ is a good choice. Next one chooses a monic polynomial $f \in \mathbf{Z}[X]$ of degree d that represents n, i.e., $f(m) = n$ for some integer m. We want both m and the coefficients of f to be as small as possible. Indeed, we can easily make them smaller than $n^{1/d}$. Now, if f is reducible, then a non-trivial factor h of f should give a non-trivial factor $h(m)$ of n. Otherwise, we consider the number field $K = \mathbf{Q}[X]/(f)$ together with the morphism φ from the order $\mathbf{Z}[\alpha]$ to $\mathbf{Z}/n\mathbf{Z}$ for which $\varphi(\alpha) = m$, where we write $\alpha = (X \bmod f)$. We then look for algebraic numbers of the form $a + b\alpha$ such that both $a + bm$ and $a + b\alpha$ are *smooth*. This means that, for a suitable positive integer y chosen at the beginning, both $a + bm$ and the norm $\mathbf{N}(a + b\alpha)$ of $a + b\alpha$ are integers divisible only by prime numbers not exceeding y. With each prime number $p \leq y$ we associate a function $\nu_p \colon \mathbf{Z} - \{0\} \to \mathbf{Z}/2\mathbf{Z}$, which assigns to any non-zero integer k the residue class modulo 2 of the number of factors p in k. In the same way, we associate with any prime ideal $\mathfrak{p}$ of $\mathbf{Z}[\alpha]$ of norm at most y the function $\nu_{\mathfrak{p}} \colon \mathbf{Z}[\alpha] - \{0\} \to \mathbf{Z}/2\mathbf{Z}$

1991 *Mathematics Subject Classification*. Primary 11Y05, 11Y40.

Key words and phrases. Factoring algorithm, algebraic number fields.

Membre de l'Option Recherche du Corps des Ingénieurs de l'Armement.

The author wishes to thank H. W. Lenstra, Jr., for his suggestions and help with the writing of this paper. Part of this work formed a D. E. A. thesis at the Ecole Polytechnique under the direction of MM. Marc Chardin, Marc Giusti and Jacques Stern, in May 1991.

that maps an element to the residue class modulo 2 of the number of factors $\mathfrak{p}$ appearing in that element (cf. [3, Section 5]).

As suggested by Adleman [1], we also use characters obtained in the following way. Choose a collection of non-zero prime ideals $\mathfrak{q}$ of $\mathbf{Z}[\alpha]$ that are different from all primes $\mathfrak{p}$ used before and that do not divide the discriminant of f. With each such $\mathfrak{q}$ we associate the function $\chi_{\mathfrak{q}} \colon \mathbf{Z}[\alpha] - \mathfrak{q} \to \mathbf{Z}/2\mathbf{Z}$ defined by

$$\chi_{\mathfrak{q}}(x) = \begin{cases} 0 & \text{if } x \text{ is a square modulo } \mathfrak{q}, \\ 1 & \text{otherwise.} \end{cases}$$

The first part of the algorithm consists of the search for many pairs (a, b) of relatively prime rational integers for which both $a + bm$ and $a + b\alpha$ are smooth, and $a + bm > 0$. In the second part one looks for subsets S of the set of pairs that have been found for which

$$\prod_{(a,b)\in S} (a + bm) \quad \text{is a square in } \mathbf{Z},$$

$$\prod_{(a,b)\in S} (a + b\alpha) \quad \text{is a square in } K.$$

One hopes that this is ensured by the following three conditions:

$$\sum_{(a,b)\in S} \nu_p(a + bm) = 0 \bmod 2 \qquad \text{for all prime numbers } p \leq y,$$

$$\sum_{(a,b)\in S} \nu_{\mathfrak{p}}(a + b\alpha) = 0 \bmod 2 \qquad \text{for all prime ideals } \mathfrak{p} \text{ of } \mathbf{Z}[\alpha] \text{ of norm } \leq y,$$

$$\sum_{(a,b)\in S} \chi_{\mathfrak{q}}(a + b\alpha) = 0 \bmod 2 \qquad \text{for all } \mathfrak{q} \text{ that have been chosen.}$$

Indeed, these conditions are necessary. If enough characters $\chi_{\mathfrak{q}}$ have been chosen then one may expect that the conditions are sufficient as well.

This leads to a large algebraic number γ that is given as a product of many small ones and that is a square in $\mathbf{Z}[\alpha]$ (see [3]):

(1) $$\gamma = f'(\alpha)^2 \cdot \prod_{(a,b)\in S} (a + b\alpha) = \beta^2 \qquad \text{with } \beta \in \mathbf{Z}[\alpha].$$

We also know that the image of γ under φ,

$$\varphi(\gamma) = f'(m)^2 \cdot \prod_{(a,b)\in S} (a + bm) \bmod n,$$

satisfies

$$f'(m)^2 \cdot \prod_{(a,b)\in S} (a + bm) = f'(m)^2 \cdot \prod_{p \leq y} p^{2e_p} = v^2,$$

where the integers e_p can be determined from the prime factorization of the numbers $a + bm$, and where

$$v = f'(m) \cdot \prod_{p \leq y} p^{e_p}.$$

As for β, we know its decomposition as an *ideal* of $\mathbf{Z}[\alpha]$; but since we do not know generators for the prime ideals of norm at most y, this does not enable us to write down an explicit expression for β itself. However, we do know an expression for the *norm* of β:

(2) $$\mathbf{N}(\beta) = \pm\mathbf{N}(f'(\alpha)) \cdot \prod_{p \leq y} p^{f_p},$$

where the f_p are non-negative integers that can be determined from the prime ideal decomposition of β. Furthermore, since $\beta \in \mathbf{Z}[\alpha]$ there exists a polynomial $B \in \mathbf{Z}[X]$ of degree at most $d - 1$ such that $\beta = B(\alpha)$. We shall compute this polynomial.

The method suggested in [3] is as follows. First, look for an odd prime q such that the polynomial f remains irreducible modulo q. Then, compute $\gamma \bmod q$ by performing all multiplications in the product (1) modulo q. We view $\gamma \bmod q$ as an element of the finite field $\mathbf{F}_{q^d}$, and we can easily compute the square roots of this element. Next, we choose one of the two square roots and lift it to a square root modulo q^2, q^4, q^8, ..., using Newton's method, until the modulus is larger than twice a given estimate of the coefficients of B. In this manner we find B. The congruence

$$B(m)^2 = \varphi(\beta)^2 = \varphi(\beta^2) = \varphi(\gamma) = v^2 \bmod n$$

suggests that $\gcd(B(m) - v, n)$ has a good chance to be a non-trivial factor of n.

One of the difficulties with this method is the very large size of the numbers that occur in the last iterations of Newton's method. The time taken by this computation is comparable to the time taken by the entire algorithm, except if one uses fast multiplication techniques; but even if that is done one may have serious practical difficulties with the very large integers that arise. It is particularly disconcerting that the huge numbers that we compute are ultimately replaced by their remainder modulo n.

The approach that we suggest is to work with many different moduli $m_i = q_i^{k_i}$, where the q_i are distinct odd primes for which f is irreducible modulo q_i. We first compute, as above, a square root β_i of γ modulo each m_i, i.e., a polynomial $B_i \in \mathbf{Z}[X]$ of degree at most $d - 1$ such that $\beta_i = B_i(\alpha)$ satisfies $\beta_i^2 \equiv \gamma \bmod m_i$; the coefficients of B_i matter only modulo m_i. If $\beta = B(\alpha)$ denotes, as above, one of the two square roots of γ in $\mathbf{Z}[\alpha]$, then we have $\beta_i = \pm\beta \bmod m_i$, where the signs are *a priori* unknown. Our first problem is to compute those signs or, equivalently, to make sure that the various β_i are congruent to the same square root β modulo m_i. Next, using the Chinese remainder theorem, we can compute $\beta = B(\alpha) \in \mathbf{Z}[\alpha]$ and $\varphi(\beta) = (B(m) \bmod n) \in \mathbf{Z}/n\mathbf{Z}$. However, the coefficients of B and the number $B(m)$ are so large that one should avoid explicitly calculating any of them. We shall see that once the $B(m) \bmod m_i$ are known, we can compute $B(m)$ modulo n without computing B or $B(m)$ itself.

2. Description and Analysis of the Method

We first consider the sign problem discussed at the end of Section 1. We shall make the assumption that the degree of the extension $K/\mathbf{Q}$ is *odd*. The basic observation is that, under this hypothesis, we have $\mathbf{N}(-x) = -\mathbf{N}(x)$ for any non-zero element x of K. Hence, exactly one of the two square roots of γ has positive norm. Let that one be called β. Suppose, as above, that we know a square root $\beta_i = B_i(\alpha)$ of γ modulo $m_i = q_i^{k_i}$ for each i, and that we want to test whether $\beta_i \equiv \beta \bmod m_i$ or $\beta_i \equiv -\beta \bmod m_i$. We can decide this by looking modulo q_i. Thus, we compute the norm of $(\beta_i \bmod q_i)$, viewed as an element of the finite field of cardinality q_i^d; this norm is the $(q_i^d - 1)/(q_i - 1)$th power of $(\beta_i \bmod q_i)$, and it belongs to the prime field $\mathbf{Z}/q_i\mathbf{Z}$. We compare this norm with the residue modulo q_i of the norm of β, which is computed by means of formula (2), but with $\pm\mathbf{N}(f'(\alpha))$ replaced by its absolute value; the multiplications in (2) are performed modulo q_i. If the two norms are equal, then $\beta_i \equiv \beta \bmod m_i$, and we keep B_i. If they are opposite, then we replace B_i by $-B_i$.

Substituting m in B_i we find $B(m)$ modulo m_i for all i, and we wish to compute $B(m)$ modulo n. We discuss this problem in a more general setting.

2.1. Changing moduli. In *modular arithmetic* one represents an integer by means of its residue classes modulo each of a set of pairwise coprime integers m_i. The theoretical basis of modular arithmetic is formed by the Chinese remainder theorem. An introduction to the algorithmic aspects of modular arithmetic, and a discussion of its applications, can be found in [4, Section 4.3.2].

We consider the following algorithmic problem from modular arithmetic (cf. [9, Section 4]). One is given a collection of pairwise coprime positive integers m_i, a positive integer n, for each i an integer x_i with $0 \le x_i < m_i$, and a small positive real number ϵ, for example $\epsilon = 0.01$. In addition, one is provided with the information that there exists an integer x satisfying $x \equiv x_i \bmod m_i$ for each i, and $|x| < (\frac{1}{2} - \epsilon)\prod_i m_i$; clearly, such an integer x is unique if it exists. The question is to compute the residue class of x modulo n.

If we define the quantities

$$
(3) \qquad\qquad M = \prod_i m_i,
$$

$$
(4) \qquad\qquad M_i = \prod_{j \ne i} m_j = M/m_i,
$$

$$
(5) \qquad a_i = 1/M_i \bmod m_i, \qquad 0 \le a_i < m_i,
$$

then the number $z = \sum_i a_i M_i x_i$ is congruent to x modulo M. Hence, if we round z/M to an integer: $r = [\frac{z}{M} + \frac{1}{2}]$, then we have $x = z - rM$. The point is that we can calculate r without calculating the possibly very large number z, as follows.

From $x = z - rM$ and our hypothesis $|x| < (\frac{1}{2} - \epsilon)M$ it follows that $\frac{z}{M} + \frac{1}{2}$ is not within ϵ of an integer. Hence, to calculate r it suffices to know an approximation t to z/M with $|t - z/M| < \epsilon$. Such an approximation can be obtained from

$$
(6) \qquad\qquad \frac{z}{M} = \sum_i \frac{a_i x_i}{m_i}.
$$

All terms in the sum are between 0 and $\max_i m_i$, so they can be computed as low precision real numbers.

This results in the following algorithm. Denote by $\mathrm{rem}(a, b)$ the remainder of the Euclidean division of a by b.

1. For each i, compute $\mathrm{rem}(M_i, m_i)$ by multiplying out the product (4) modulo m_i, and compute the numbers a_i as in (5) with the extended Euclidean algorithm.

2. Compute $\mathrm{rem}(M, n)$ by multiplying out the product (3) modulo n, and compute $\mathrm{rem}(M_i, n)$ for each i; if $\gcd(m_i, n) = 1$ one can do this by dividing $\mathrm{rem}(M, n)$ by m_i modulo n;

3. Compute a number t that differs by less than ϵ from the sum in (6), and round t to an integer: $r = [t + \frac{1}{2}]$.

4. Output

$$\mathrm{rem}(x, n) = \mathrm{rem}\left(\left(\sum_i a_i \,\mathrm{rem}(M_i, n)x_i\right) - r\,\mathrm{rem}(M, n), n\right).$$

(If m_i is much larger than n one may prefer to replace a_i and x_i by $\mathrm{rem}(a_i, n)$ and $\mathrm{rem}(x_i, n)$ in this expression.)

Note that we never handle numbers substantially larger than the moduli.

2.2. Size and complexity. We derive upper bounds for the integers b_i for which $\beta = b_0 + b_1\alpha + \cdots + b_{d-1}\alpha^{d-1}$, with β as in (1). For $f = \sum_{i=0}^d c_i X^i$ we write $\|f\| = \left(\sum_{i=0}^d c_i^2\right)^{1/2}$, and we let u be an upper bound for all numbers $|a|$, $|b|$ for which $(a, b) \in S$.

Proposition. *We have*

$$|b_i| \le d^{3/2} \cdot \|f\|^{d-i} \cdot (2u\|f\|)^{\#S/2}$$

for $0 \le i \le d-1$.

Proof. The field K has d embeddings into the field of complex numbers, and we denote the image of an element $\epsilon \in K$ under the kth embedding by $\epsilon^{(k)}$. We have $f = \prod_{i=1}^d (X - \alpha^{(i)})$, and

$$(7) \qquad \max(1, |\alpha^{(k)}|) \le \prod_{j=1}^d \max(1, |\alpha^{(j)}|) \le \|f\|$$

for each k; the first inequality is trivial, and the second is due to Landau [5; 8, Chapitre IV, Section 3.3].

Let $\delta_0, \delta_1, \ldots, \delta_{d-1} \in K$ be defined by $\sum_{i=0}^{d-1} \delta_i X^i = f/(X - \alpha)$, so that $\delta_i = \sum_{j=0}^{d-1-i} c_{i+j+1}\alpha^j$. By [6, Chapter III, Proposition 2] we have

$$(8) \qquad b_i = \mathrm{Tr}(\delta_i\beta/f'(\alpha)) = \sum_{k=1}^d \delta_i^{(k)}\beta^{(k)}/f'(\alpha^{(k)}),$$

where $\mathrm{Tr}\colon K \to \mathbf{Q}$ is the trace function. The Cauchy-Schwarz inequality and (7) imply that

$$|\delta_i^{(k)}|^2 \leq \|f\|^2 \cdot \sum_{j=0}^{d-1-i} |\alpha^{(k)}|^{2j} \leq d \cdot \|f\|^{2(d-i)} \qquad (0 \leq i \leq d-1).$$

From

$$|\beta^{(k)}/f'(\alpha^{(k)})|^2 = \prod_{(a,b)\in S} |a + b\alpha^{(k)}| \leq (2u\|f\|)^{\#S}$$

we now obtain

$$|b_i| = \Big|\sum_{k=1}^{d} \delta_i^{(k)}\beta^{(k)}/f'(\alpha^{(k)})\Big| \leq d^{3/2} \cdot \|f\|^{d-i} \cdot (2u\|f\|)^{\#S/2},$$

as required.

If f is chosen as in [3], then we have $\|f\| \leq \sqrt{d} \cdot n^{1/d}$ and $|m| \leq n^{1/d}$, and therefore

$$(9) \qquad\qquad |B(m)| \leq d^{(d+5)/2} \cdot n \cdot (2u\sqrt{d}n^{1/d})^{\#S/2}.$$

Note that the three factors in this upper bound are of completely different orders of magnitude. In realistic cases, the last factor has millions of decimal digits, the middle one has between one and two hundred digits, and the first one has just a few digits. We refer to [3, 9.3 and Section 11] for an estimate of u and $\#S$ as functions of n and d.

The estimate (9) is good enough to enable us to get a rough impression of the precision needed, i.e., the number and size of moduli m_i. Note that too many moduli would make us waste time, while not enough moduli would give an incorrect result. In order to get a more accurate estimate, we can explicitly compute the zeroes $\alpha^{(k)}$ of f as low precision complex numbers. We can then evaluate the complex numbers $\delta_i^{(k)}$ and $f'(\alpha^{(k)})$, and, multiplying out the product (1), the real numbers $|\beta^{(k)}|$. Then we obtain from (8) an upper bound for $|b_i|$. This leads to an upper bound for $B(m)$ that is better than (9).

We now give a rough estimate of the complexity of our method as a function of the logarithm of an upper bound like (9) for $B(m)$; let this logarithm be called s. For the sake of comparison, we mention that the square root method proposed in [3] takes time $s^{1+o(1)}$ if fast multiplication techniques are used, and $s^{2+o(1)}$ otherwise, and that the entire number field sieve is conjectured to run in time $s^{2+o(1)}$, see [3, 9.3 and Section 11]; here and below the $o(1)$ is for $n \to \infty$. The time taken by our method depends on the size and the number of the moduli, and on whether or not fast multiplication techniques are used. We consider two extreme cases.

In the first case the moduli are chosen as small as possible. We assume, heuristically, that f is irreducible modulo one out of every d primes (see the remark below). We take the m_i to be all those primes up to $ds(1 + o(1))$. Their product will then be $\exp(s^{1+o(1)})$, which is large enough for the algorithm of 2.1. From $d = s^{o(1)}$ one sees that both the number of moduli and the largest of them is of the order $s^{1+o(1)}$. The most time-consuming part of the method is the computation of the product (1) modulo each m_i. This requires $s^{2+o(1)}$ multiplications. Since all multiplications are done with small numbers, it is not clear how fast multiplication techniques can help at all. It is conceivable, though, that the computation can be speeded up by some divide-and-conquer technique.

The second extreme possibility is to fix the number of moduli and to choose each m_i to be $\exp(s^{1+o(1)})$. In this case we can use fast multiplication, and the time taken is $s^{1+o(1)}$. This is also achieved in [3] by means of a single modulus. With this choice of moduli one does ultimately handle very large integers.

The conclusion is that from a theoretical point of view our method does not represent an improvement over [3]. In practice, however, our method has the advantage of offering the possibility to work with much smaller numbers, and in addition it can be run in parallel. The number and the size of the moduli to be used depend strongly on the features of the available computing equipment. In [2] many small moduli are chosen, which is justified by the use of a massively parallel computer. One can imagine that in other situations it is desirable to use larger moduli so as to take advantage of sophisticated multiplication techniques. In principle it is not necessary to let the moduli be of the same approximate size: if several different computers are used, then one can adapt the sizes of the moduli to the individual machines.

Remark. It was pointed out in [3] that primes q for which f is irreducible modulo q do not necessarily exist, but that for "most" f one may expect that one out of every d primes has this property (see [10]). If the degree d, which we assumed to be odd, is a prime number, then indeed every irreducible polynomial $f \in \mathbf{Z}[X]$ of degree d remains irreducible modulo at least one out of every d primes (asymptotically). This applies in particular to $d = 3$, 5, and 7. For the proof it suffices, by the argument of [3, Proposition 9.1], to show that every transitive permutation group G of prime degree d contains at least $\#G/d$ cycles of length d. Indeed, by Cauchy's theorem G has an element of order d, and this element must be a d-cycle. Letting G act by conjugation on the set of its d-cycles one finds that the number of d-cycles is divisible by $\#G/d$, as required.

The complete algorithm may be summarized as follows.

Algorithm. Let the integers n and m, the integer $v \bmod n$, the polynomial f, and the set S of pairs (a, b) be given, as in the number field sieve. This algorithm determines the possibly trivial factorization of n that the set S gives rise to.

1. Choose the moduli $m_i = q_i^{k_i}$ according to the above remarks. This necessitates the computation of complex approximations to the $\alpha^{(i)}$ and $|\beta^{(i)}|$.
2. Compute for each i the numbers $\mathrm{rem}(M_i, m_i)$, a_i, and $\mathrm{rem}(M_i, n)$ as in 2.1.
3. For each modulus m_i, compute the product

$$\gamma_i = f'^2 \cdot \prod_{(a,b)\in S} (a + bX) \bmod (f, m_i),$$

as well as a square root β_i of γ_i.

4. Compute $N_{1,i}$, the norm of β_i mod q_i, and $N_{2,i}$, the product in (2) modulo q_i. If $N_{1,i} \neq N_{2,i}$ then replace β_i by $-\beta_i$. Let $B_i \in \mathbf{Z}[X]$ be a polynomial of degree $\leq d - 1$ for which $\beta_i = B_i$ mod (f, m_i), and compute $B_i(m)$ (modulo m_i). This step is to be performed for each i.

5. Compute $B(m)$ mod n from the $B_i(m)$ using the quantities calculated in step 2 (see 2.1).

6. Output $\gcd(B(m) - v, n)$.

Note that steps 2, 3, and 4 can be parallelized.

References

1. L. M. Adleman, *Factoring numbers using singular integers*, Proc. 23rd Annual ACM Symp. on Theory of Computing (STOC) (1991), 64–71.

2. D. J. Bernstein, A. K. Lenstra, *A general number field sieve implementation*, this volume, pp. 103–126.

3. J. P. Buhler, H. W. Lenstra, Jr., Carl Pomerance, *Factoring integers with the number field sieve*, this volume, pp. 50–94.

4. D. E. Knuth, *The art of computer programming*, volume 2, second edition, Addison-Wesley, Reading, Mass., 1981.

5. E. Landau, *Sur quelques théorèmes de M. Petrovič relatifs aux zéros des fonctions analytiques*, Bull. Soc. Math. France **33** (1905), 251–261.

6. S. Lang, *Algebraic number theory*, Addison-Wesley, Reading, Massachusetts, 1970.

7. A. K. Lenstra, H. W. Lenstra, Jr., M. S. Manasse, J. M. Pollard, *The number field sieve*, this volume, pp. 11–42. Extended abstract: Proc. 22nd Annual ACM Symp. on Theory of Computing (STOC) (1990), 564–572.

8. M. Mignotte, *Mathématiques pour le calcul formel*, Presses Universitaires de France, Paris, 1989.

9. P. L. Montgomery, R. D. Silverman, *An FFT extension to the $P - 1$ factoring algorithm*, Math. Comp. **54** (1990), 839–854.

10. B. L. van der Waerden, *Algebra*, seventh edition, Springer-Verlag, Berlin, 1966.

U. M. R. D'Algorithmique Arithmétique de Bordeaux, Université de Bordeaux I

Groupe de recherche en complexité et cryptographie, L. I. E. N. S., URA 1327 du CNRS, D. M. I., Ecole Normale Supérieure, 45 rue d'Ulm, 75230 Paris Cedex 05, France

E-mail address: couveign@dmi.ens.fr

A GENERAL NUMBER FIELD SIEVE IMPLEMENTATION

DANIEL J. BERNSTEIN, A. K. LENSTRA

ABSTRACT. The general number field sieve is the asymptotically fastest—and by far most complex—factoring algorithm known. We have implemented this algorithm, including five practical improvements: projective polynomials, the lattice sieve, the large prime variation, character columns, and the positive square root method. In this paper we describe our implementation and list some factorizations we obtained, including the record factorization of $2^{523} - 1$.

1. INTRODUCTION

The general number field sieve (GNFS) [3] is a modified version of the special number field sieve (SNFS) [8; 9]. GNFS factors arbitrary integers n in heuristic time

$$\exp((c_g + o(1)) \log^{1/3} n \log^{2/3} \log n)$$

as $n \to \infty$. Here $c_g = (64/9)^{1/3} \approx 1.9$. For the special integers n where SNFS is applicable, GNFS takes time

$$\exp((c_s + o(1)) \log^{1/3} n \log^{2/3} \log n),$$

with $c_s = (32/9)^{1/3} \approx 1.5$. These asymptotic estimates should be compared with the time

$$\exp((1 + o(1)) \log^{1/2} n \log^{1/2} \log n)$$

taken by the multiple polynomial quadratic sieve (QS) [14], generally regarded as the best general-purpose factoring method for large integers.

We implemented GNFS on Bellcore's MasPar[1], a SIMD (single instruction multiple data) computer with 16384 processors, and used it to obtain some

1991 *Mathematics Subject Classification*. Primary 11Y05, 11Y40.

Key words and phrases. Factoring algorithm, algebraic number fields.

Thanks to Joe Buhler, Hendrik Lenstra, John Pollard, and Carl Pomerance for their helpful suggestions, and to Andrew Odlyzko for his help with the factorization of $2^{523} - 1$. The first author was supported in part by a National Science Foundation Graduate Fellowship and by Bellcore.

[1] It is the policy of Bellcore to avoid any statements of comparative analysis or evaluation of products or vendors. Any mention of products or vendors in this presentation or accompanying printed materials is done where necessary for the sake of scientific accuracy and precision, or to provide an example of a technology for illustrative purposes, and should not be construed as either a positive or negative commentary on that product or vendor. Neither the inclusion of a product or a vendor in this presentation or accompanying printed materials, nor the omission of a product or a vendor, should be interpreted as indicating a position or opinion of that product or vendor on the part of the presenter or Bellcore.

factorizations, as described in Section 12. To speed the implementation we used five modifications to the basic algorithm: projective polynomials [3], Pollard's lattice sieve [13], the large prime variation [8; 9; 10], character columns [1], and Couveignes's positive square root method [4]. In this paper we describe our implementation. We also present some factorizations we obtained with GNFS.

We suspect that for general numbers GNFS is competitive with QS at the edge of what we can factor in a reasonable time today, between 120 and 130 digits. For larger numbers it is faster. For smaller numbers it is usable, though slower than QS in its current form—but see Section 13 for comments on this situation.

We warn the reader that the word "smooth" is defined anew in several sections of this paper. To avoid confusion we give each different use of "smooth" a unique prefix: basically smooth in Section 3, sieve-smooth and prec-smooth in Section 6, and rat-smooth and alg-smooth in Section 7. We write $\#S$ for the size of a set S. Most further notation for this paper is established in Section 3.

2. OUTLINE OF THE IMPLEMENTATION

We refer the reader to [3] for an explanation of how the number field sieve works; except for a few brief proofs in Section 11 we do not justify the algorithm here.

We begin with a composite number $n > 0$ to be factored. We assume as known that n is odd—in fact that it has no factors smaller than some reasonable bound, say the largest integer that fits into a computer word. We also assume that n is not a power of a prime. These facts will all be apparent after an application of trial division, standard compositeness and power-of-prime tests [8], Pollard's ρ method [7], and the elliptic curve method [11], all of which are tried before the number field sieve.

GNFS can be divided into four stages. In the first stage, described in Sections 3 through 5, we choose parameters and precompute various information about an algebraic number field associated with n. In the second and most time-consuming stage, described in Sections 6 and 7, we perform the heart of the number field sieve, looking for "relations." In the third stage, described in Sections 8 and 9, we build a matrix out of the relations and find random elements of the nullspace of the matrix. In the final stage, described in Sections 10 and 11, we effectively take a square root of a large element of our number field.

Figure 1 shows the flow of data in GNFS. Upper case names such as **POLY** are data files, i.e., sets stored on the computer; lower case names such as **pnfls** are programs. Each node is labeled with a section number. Data flows downwards through the graph.

3. CHOOSING THE POLYNOMIAL

The entire computation begins with a file **POLY**, which contains the number n to be factored together with a low-degree polynomial f and an integer m such that n divides $f(m)$. The degree d of f is most commonly 3 or 5; it is required to be odd because of a restriction in the square root stage, as described in Section 10.

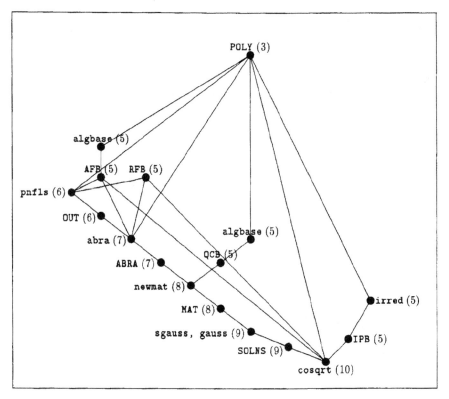

FIGURE 1.   The flow of data in our GNFS implementation.

(See Remark 10.1. We can in principle use even degrees, as we have a single-processor implementation of the square root method described in [3], which can work with any degree. We have been using a MasPar implementation of another method, which demands odd degrees.)

Several mathematical objects are implicitly associated with the polynomial f we choose. Most important is the projective two-variable norm polynomial, which we define as $N(a, b) = f(a/b)b^d$; see Section 11 for a number-theoretic interpretation of this object. Another is the monic polynomial $g(x) = f(x/c_d)c_d^{d-1}$, where c_d is the leading coefficient of f. Another is the number field

$$\mathbf{Q}[\alpha]/f(\alpha) = \mathbf{Q}[\omega]/g(\omega).$$

Here α could be regarded as a root of f within the complex plane, and $\omega = \alpha c_d$ could be regarded as a root of g; but neither α nor ω is ever computed explicitly. We actually work within the number ring $\mathbf{Z}[\omega]/g(\omega)$ inside our number field. Elements of this number ring are represented as polynomials $\sum_{0 \leq i < d} z_i \omega^i$ where each z_i is an integer. Finally we will need the discriminant $\operatorname{disc}(g)$ (but see Remark 10.4 for a way to avoid computing this quantity).

For special numbers n, such as numbers of the form $b^c \pm 1$ where SNFS is applicable, we choose POLY by hand. Otherwise we let the computer search for a good polynomial by brute force in a given range of m's, as described in the next section.

All else being equal, one polynomial f is better than another with a similar value of m if the norm values $N(a, b)$ are more likely to be basically smooth for a and b small. Here x is basically smooth if its prime factors are small. (It is not necessary to define "small" precisely.) See next section for further details of how we find f in general.

We measure f (for a fixed d) in two ways. The first is coefficient size: if f has coefficients that are small in absolute value then $N(a, b)$ will be small and hence likely to be basically smooth. The second measurement is more algebraic: we compute an average logarithmic subtraction from $\log N(a, b)$ given the roots of f modulo tiny primes. Consider, for example, a polynomial f congruent to $x^4 - 1$ modulo 5, so that f has four roots modulo 5. Then $N(a, b)$ will be divisible by 5 for four-fifths of all choices of a, b, in some sense. So if we remove the first factor of 5 (if any) from $N(a, b)$ then $\log N(a, b)$ becomes an average of $(4/5) \log 5$ smaller. We compute these logarithmic subtractions for tiny primes through 19.

Once we have chosen f we choose several further parameters: the rational prime bound max RFB (on the order of 10^6), the algebraic prime bound max AFB (so far always chosen equal to max RFB), the large prime bound L (on the order of 10^8), a quadratic character base size #QCB (so far always chosen as 100), an inert prime base size #IPB (on the order of 10^5), and a range $\{e\}$ of integers (typically all values between 1 and 10^4—see Remark 3.1).

All these parameters except #IPB and #QCB are chosen with the constraint that the matrix constructed in Section 8 has a nontrivial nullspace. The size #QCB must be large enough that the quadratic characters are highly likely to span a certain vector space; see Section 8 and [3]. The size #IPB must be large enough that a certain inequality holds; see Remark 10.3 and Remark 11.1.

3.1. *Remark.* The range $\{e\}$ determines the length of one of the sides of the sieving parallelogram (see Section 6), and can be chosen freely. The other length is hard-wired in our sieving program and fixed at compile time; its choice depends on the design of the sieving program and is severely limited by the architecture of the MasPar.

3.2. *Remark.* In principle we should check that m and c_d do not have any factors in common with n, that the derivative $f'(m)$ does not have any factors in common with n, and that there are enough inert primes, but in practice none of these problems will occur. If f is "bad" in any way then the algorithm will fail in an obvious manner. We describe these failures at the points where they may occur. See Remark 5.2 and Remark 10.2.

3.3. *Remark.* The possibility of f not being monic characterizes the "projective" number field sieve. It permits much smaller coefficients at very little expense. Actually the projective number field can be chosen with two parameters m_1 and m_2 in place of m; the idea is that m is a fraction, m_1/m_2. The additional possibility of $m_2 \neq 1$ characterizes the "homogeneous" polynomial variation. See [3] for details of this generalization. As m_2 intrudes (a bit) at every step of the computation, and does not show any obvious advantages, we have not used it. But see Section 13 for further comments.

4. SEARCHING FOR A POLYNOMIAL

The time taken by GNFS depends heavily on the size of the polynomial f. In this section we describe how we search for polynomials when we do not know a special form for n.

To fix ideas we set $d = 5$. Here is the polynomial-searching problem. We are given n. We want to find a positive integer m in a given range and six integer coefficients $c_0, \ldots, c_5$ such that $c_0 + c_1 m + \cdots + c_5 m^5$ is a positive multiple of n, with all the c_i small in absolute value.

The straightforward solution is to pick m around $n^{1/6}$ and to expand n in base m. But there are many other polynomials for n. For a detailed analysis of the polynomial existence problem we refer to [3]; here is a rough outline of the argument showing that we can expect that much better polynomials exist. Say $d = 5$ and n is around 10^{145}. We have seven parameters, m and c_0 through c_5, to change; let us choose m around 25 digits, c_5 around 20 digits, and the other c_i up to 25 digits. This gives approximately 10^{170} values of $f(m)$; it is reasonable to expect that at least 10^{25} of them will equal (or at least be divisible by) n. The five c_i other than c_5 vary wildly with the precise value of m. If they are independently and uniformly distributed then there is some polynomial for n where each of those five c_i is a factor of $10^{25/5} = 10^5$ better than the maximum. In other words there should be a polynomial for n with $m \approx 10^{25}$ and $|c_i| < 10^{20}$.

By the same argument, if we use $m = m_1/m_2$ as described in [3], choosing m_1 and m_2 around 25 digits, then there should be a polynomial for n with $m_j \approx 10^{25}$ and $|c_i| < 10^{16}$. But it is already extremely difficult to find optimal c_i without m_2, so we will ignore m_2 for the remainder of this section. In case of any breakthrough in polynomial-searching methods it will obviously be important to use m_2.

Unfortunately we do not know any way substantially faster than brute force to generate a good polynomial for general n. We start from a polynomial with a given m. We choose some integer k, replace m by $m - k$, and compute the polynomial for this new value of m. If any coefficient is larger than $m/2$ in absolute value, we subtract m from it and add 1 to the next coefficient. Finally we check whether all the coefficients are small. We repeat as long as necessary.

It is easy to choose k so that c_4 is bounded by a small multiple of c_5. But this still leaves four coefficients, c_0 through c_3, which can range up to about $m/2$. If we search through 10^{12} polynomials in this way then we expect to find some value of m such that each of c_0 through c_3 is at most $m/2000$ in absolute value. This is good, though still nowhere near optimal.

We emphasize that polynomial-searching is highly underdeveloped. There is much unexploited structure in the polynomial-searching problem. It *appears* far more tractable than factoring itself. Surely we can do better than brute force?

5. COMPUTING BASES

After settling on a polynomial we compute four bases: the rational factor base RFB, the algebraic factor base AFB, the quadratic character base QCB, and the inert prime base IPB. All these files are stored in binary format so that they can be loaded quickly. All the parameters listed in Section 3 except {e} are used in constructing these factor bases.

RFB does not depend on POLY. It consists of all odd primes up to maxRFB.

AFB is a list of all (p, r) pairs with p dividing $f(r)$ and $0 \leq r < p$; p is restricted to the odd primes below maxAFB. There may be as many as d roots r of f modulo p. Our algbase program computes them as in [7, Section 4.6.2]. We also include pairs (p, ∞) (represented inside the computer as (p, p)) with p dividing c_d, as per [3]. For all polynomials used in practice, the number of (p, r) pairs in the algebraic factor base with p in any given range will be very close to the number of primes p in the rational factor base in the same range.

IPB is a list of #IPB odd primes ℓ that remain inert in the number ring, i.e., for which f is irreducible modulo ℓ. We also require that no ℓ divide c_d. These conditions imply that ℓ does not divide the discriminant disc(g). See Remark 11.1 for details on the choice of #IPB. A program irred, similar to algbase but running on the MasPar, looks for inert primes among all primes $\ell > 10000$. (Actually it would be better to generate primes ℓ down from the computer's word size; see Remark 11.1.) It checks 16384 possibilities for ℓ at once using the tests in [7, Section 4.6.2].

Finally, QCB is a list of (q, s) pairs with the odd prime q dividing $f(s)$; here, unlike AFB, q is required to be larger than the large prime bound L. We choose the first #QCB primes q larger than L. As indicated in Section 3 we use only #QCB = 100 different pairs, so QCB is a small file.

5.1. *Remark.* In principle the q's in QCB should not divide c_d or the derivative $f'(s)$. These conditions rarely fail, but if they do fail then they will fail silently (i.e., without overt effects on the computation). On the other hand, we have not bothered checking for these failures, for reasons explained in Remark 8.1.

5.2. *Remark.* The files computed here could, in highly improbable circumstances, be much too small. In this case f is bad and a different f must be chosen.

6. SIEVING

In this section we regard the rational primes p as pairs $(p, m \bmod p)$. We thus regard the rational and algebraic factor bases as a single collection of pairs (p, r), of size #RFB + #AFB. (For simplicity we ignore any pairs (p, ∞).) In this section we use the parameter {e} as well as many sieving parameters chosen heuristically.

The object of the sieving stage is to produce an (a, b) pair, with a and b relatively prime integers that fit easily inside a computer word, such that $b > 0$ and $a \equiv br \pmod{p}$ for several pairs (p, r)—a sieve-smooth (a, b) pair, for short. Actually the sieving stage should produce as many sieve-smooth (a, b) pairs as possible in a reasonable amount of time.

6.1. *Sieving without special-q.* Here is the direct approach to sieving. We consider a region of the (a, b)-plane of area A. We divide the region into some number X of roughly equal-area pieces. We then do the following:

1.   For each piece
2.   For each (p, r)
3.   Figure out which (a, b) inside the piece have $a \equiv br$ (mod p)
4.   "Hit" all those spots
5.   For each (a, b) in the piece
6.   Check if (a, b) has enough hits

For example, one might choose $X = 1$ and initialize an array of counters, one counter for each (a, b) pair in a rectangle near the origin, say $0 < b < v$ and $-u < a < u$. (With these choices we say we are using the "plane method.") For each pair (p, r) one "hits" the counters for all (a, b) pairs in the lattice defined by $a \equiv br$ (mod p). After all (p, r) have been considered one can "read off" the sieve-smooth (a, b) pairs from the counters.

As another example, in other NFS implementations [8; 9] one runs through the (a, b) plane in horizontal slices. More precisely, one still considers $0 < b < v$ and $-u < a < u$, but this area is split into $X = v - 1$ pieces, each piece with a fixed b. (This is the "line method.") For each $b = 1, 2, \ldots$ one initializes counters for the line of a values between $-u$ and u. For each pair (p, r) one runs through the a values with $a \equiv br$ (mod p), hitting the appropriate counters. After this one can read off the sieve-smooth (a, b) pairs and continue on to the next b. One feature of this method is that it is easy to keep track of br (mod p) as b increases. More importantly, we end up using only a small amount of memory.

As yet another example, one can split the plane into horizontal slices, and then split each slice into several pieces. This reduces memory requirements still further.

How much time and memory does the direct approach take? Denote the set of (p, r) by P. We use time $T_1 X$ for step 1, time $T_2 X \# P$ for steps 2 and 3, and time $T_4 A$ for steps 5 and 6; recall that A is the total area of the (a, b) plane under consideration. (Here each T_i is the amount of time taken by some basic step; see Remark 6.13.) We use time roughly $T_3 \sum_{(p,r) \in P} A/p$ for step 4, assuming (as is correct for current theory and practice) that the A pairs (a, b) fall almost equally into equivalence classes of $a - br$ modulo p.

The total time is

$$(6.2) \qquad T_1 X + T_2 X \# P + T_4 A + T_3 A \sum 1/p.$$

The total memory is some constant plus A/X.

6.3. *Sieving with special-q.* In the direct approach described above, we must consider every pair (a, b). Pollard [13] pointed out that a variation of the "special-q" method for QS of Davis and Holdridge [5] could be applied to NFS. This variation, called the "lattice sieve" for NFS by Pollard, is a general improvement, which with the proper parameter selection lets us consider only a fraction of all (a, b) pairs.

Here is how it works. Choose an ordering of the set P of pairs (p, r)—see Remark 6.9. For each (a, b) define "the special (q, s)" as the largest (p, r) with $a \equiv br \pmod{p}$. Then partition all (a, b) pairs according to their special (q, s) values. Instead of searching through all (a, b) we search through the (a, b) for each special $(q, s) \in Q$ in turn, for some subset Q of P. Some (a, b) pairs don't have a special (q, s) in Q, and thus are ignored, but with the proper choice of Q we can find most sieve-smooth (a, b) pairs.

Fix $(q, s) \in Q$. The set of sieve-smooth (a, b) pairs with this special (q, s) is a subset of the lattice of (a, b) pairs with $a \equiv bs \pmod{q}$. Which subset is it? It is the subset that is sieve-smooth for those (p, r) pairs no larger than (q, s).

So we run through all $(q, s) \in Q$ in turn. For a fixed (q, s) we compute two short vectors $\mathbf{C} = (a_c, b_c)$ and $\mathbf{E} = (a_e, b_e)$ that generate the lattice $a \equiv bs \pmod{q}$. (If we cannot find short enough vectors we simply throw q away. See Remark 6.11.) Then we find pairs (c, e) such that $(a, b) = c\mathbf{C} + e\mathbf{E}$ is hit by several (p, r) smaller than (q, s). To do this we transform the condition "$a \equiv br \pmod{p}$" into a similar condition on (c, e), and we search through a small rectangle of (c, e) values near the origin.

More formally, here is the algorithm.

1.   For each (q, s) in Q
2.   Identify the lattice of (a, b) with $a \equiv bs \pmod{q}$
3.   Form coordinates (c, e) for some area of size C of this lattice
4.   For each (p, r) in P
5.   Transform "$a \equiv br \pmod{p}$" to the (c, e) plane
6.   For each piece of the (c, e) plane
7.   For each (p, r) smaller than (q, s)
8.   "Hit" the appropriate (c, e) inside the piece
9.   For each (c, e) in the piece
10.   Check if (c, e) has enough hits

We have two parameters here corresponding to the parameters A and X in the direct approach: we choose an area of size C in the (c, e)-plane, and split it into Y roughly equal pieces for steps 6 through 10.

For example we can choose $Y = 1$ and use the plane method ("sieving by vectors") in the (c, e) plane. Or we can split the (c, e) plane down into $Y > 1$ lines, each with a fixed c—hence using the line method ("sieving by rows") in the (c, e) plane. We can even split the (c, e) plane into pieces smaller than lines, so as to save on memory. We could choose a different region of the (c, e)-plane for each (q, s) (as suggested in [13]), but for simplicity we keep C and Y constant (also following [13]).

How much time and memory does the special-q approach take? To simplify the analysis let us assume (to the detriment of the lattice sieve) that all pairs (p, r), rather than just those pairs up through (q, s), are used in steps 7 through 10.

As in our analysis of the direct approach, for each (q, s) we do a sieve taking time $T_1 Y + T_2 Y \#P + T_4 C + T_3 C \sum_{(p,r)} 1/p$. The total over all (q, s) is $T_1 Y \#Q + T_2 Y \#P \#Q + T_4 C \#Q + T_3 C \#Q \sum 1/p$. We also spend time $T_5 \#Q + T_6 \#P \#Q$

on the preparatory steps of the lattice sieve. The total time is

$$(6.4) \qquad \#Q\big(T_1 Y + T_2 Y \#P + T_4 C + T_3 C \sum 1/p + T_5 + T_6 \#P\big).$$

The total memory is some constant plus C/Y.

6.5. *Comparison of sieving methods.* We have presented two different approaches: the direct approach and the special-q approach. Each approach uses parameters that determine the time, memory, and yield of the sieving step: the direct approach uses X pieces of total area A in the (a, b) plane, and the special-q approach uses Y pieces of total area C in the (c, e) plane.

To compare the approaches we must first assume that they give the same yield. An area of A pairs (a, b) corresponds roughly to an area of $A/\max P$ of the (c, e) plane; as illustrated by Pollard [13] we will miss some, though not too many, sieve-smooth (a, b) pairs if we set C as small as $A/\max P$. To equalize the yield we might choose an area as large as $10 A/\max P$ of the (c, e) plane. This number 10 is surely large enough if we choose Q following Pollard. In the following discussion we will take $C = V A/\max P$ where V is an unknown "fuzz factor" between 1 and 10.

Now for fixed A and C our time analyses (6.2), (6.4) show that sieving is faster when X and Y are smaller—more pieces means more overhead. On the other hand we use A/X or C/Y memory, so if memory is limited then we cannot choose X or Y too small.

Therefore, no matter which approach we use, we split the plane into as few pieces as possible, subject to the sole constraint that our sieve fit into memory. This is important both for theory and practice.·

To continue our comparison of the two approaches we must consider two cases. Either we have an incredibly large amount of memory available (at least $A/\max P$), or we do not. In the first case we can choose $Y = 1$ (and $X \leq \max P$). Again to the detriment of the lattice sieve we will ignore the X terms in (6.2) altogether: now the time (6.2) for the direct approach is over $A(T_4 + T_3 \sum 1/p)$, and the time (6.4) for the special-q approach is

$$A \frac{\#Q}{\max P} \left((T_1 + T_5) \frac{\max P}{A} + (T_2 + T_6) \frac{\#P \max P}{A} + T_4 V + T_3 V \sum \frac{1}{p} \right).$$

Both in theory [3, Section 10] and in practice we may assume that A is at least roughly as large as $\#P \max P$. Now it is easy to see that the special-q approach is asymptotically some constant times $\max P/\#Q > \log \max P$ faster than the direct approach.

In the second case we do not have $A/\max P$ memory available. Then both approaches are constrained by the available memory, so as per our previous comments we assume that they both use as much memory as possible—in particular, they both use the same amount of memory. In other words A/X and $C/Y = V A/(Y \max P)$ should be equal. Hence $Y = X V/\max P$.

Now the time (6.2) for the direct approach is

$$X(T_1 + T_2 \#P) + A(T_4 + T_3 \sum \frac{1}{p}),$$

and the time (6.4) for the special-q approach is

$$X \frac{V \# Q}{\max P}(T_1 + T_2 \# P) + A \frac{V \# Q}{\max P}\left(T_4 + T_3 \sum \frac{1}{p} + T_5 \frac{\max P}{VA} + T_6 \frac{\# P \max P}{VA}\right).$$

The first terms may dominate the second terms, or they might not. In either case, the special-q approach is once again asymptotically $\log \max P$ faster than the direct approach.

Of course, the special-q approach is more complex than the direct approach, so in small examples [13] the direct approach may be faster, depending on the constants T_i and V. But as $\max P$ grows the special-q approach has a $\log \max P$ advantage, both in theory and in practice. For us $\max P$ is large enough that we suspect the special-q method is more than twice as fast as any implementation of the direct approach.

6.6. *Sieving on a MasPar.* The MasPar consists of a mesh of 128×128 SIMD processors with 64 KByte memory per processor. As shown in [6] the multiple polynomial variant of the quadratic sieve can be implemented quite efficiently on the MasPar. The MasPar is split into 128 rows, each handling the sieve line for one polynomial. Each sieve line is split into 128 pieces, one piece per processor. We use just 32 KByte on each processor to avoid conflicts with other users; of this we allocate 27000 bytes for the sieve.

To implement the sieving stage of NFS on the MasPar we could follow the same approach—i.e., the direct approach, with $A/X = 27000$ values of (a, b) per piece, and one piece per processor. However, we never attempted to implement this. Instead we tried to get the lattice sieve—special-q, with the plane method in the (c, e) plane—to work on the MasPar. Unfortunately we have not come up with a way to lay out the (c, e) plane over more than one processor without running into serious communications overhead.

So we chop the (c, e) plane into pieces that fit into one processor. In Pollard's terminology [13] we "sieve by rows": each processor considers one e for one (q, s) and uses its memory to sieve a line of c's with all the pairs (p, r) smaller than (q, s). Our MasPar sieving program, pnfls, handles 128 special (q, s) pairs at a time, one pair per row of processors. For each (q, s) batch it iterates through 128 e's from the range $\{e\}$ at a time, one e per column of processors.

6.7. *Sieving results.* The sieving program pnfls produces some "reports" of good (c, e) pairs, which (together with the corresponding (a, b) pairs and q's) are saved in a binary file OUT.

Our first version of pnfls used 10^4 values of c, namely $-5000 \le c < 5000$, and used 24 KByte per processor. We used counters of the form (x_r, x_a), initialized to $(0, 0)$ and stored in two bytes. Denote the approximate logarithm base 2 of p by $l(p)$. To hit a counter with a prime p from RFB we added $(l(p), 0)$ to the counter. To hit a counter with a prime pair (p, r) from AFB we added $(0, l(p))$ to the counter. We choose, by hand, minimum values for x_r and x_a; we throw out any (c, e) whose counter does not meet or exceed those minima. Pairs (c, e) that pass this test are subjected to more stringent tests, involving approximations to

$a - bm$ and $N(a, b)$. When all is said and done perhaps one out of every 7 or 8 reported sieve-smooth (a, b) pairs is actually prec-smooth (precisely smooth). (See Remark 6.8.) Hopefully we do not throw out more than a fraction of prec-smooth pairs.

In the current version of **pnfls** we use the same byte to represent x_r during the first sieve and x_a during the second one: we first sieve with the primes from **RFB**, remember the locations of x_r's where the counter is large enough before resetting the x_r's to zero, and next use the same sieve locations to sieve with the pairs from **AFB**. In this way we could double the length of the c interval and increase the yield per unit of time, without affecting the memory requirements of **pnfls**. With a c interval of $-10000 \leq c < 10000$, **pnfls** fits in 24 KByte per processor; currently we are using $-14000 \leq c < 14000$, which fits in 32 KByte.

6.8. *Remark.* In the above description we neglected to define various terms: "several pairs," "hitting," "read off," "most," "small," etc. The real object of the sieve stage is to limit the number of (a, b) pairs that have to be checked by trial division (next section). For trial division there is a quite precise definition of what "smoothness" (prec-smoothness) means: namely, $a - bm$ factors almost completely into primes smaller than max **RFB** with at most one "large prime" between max **RFB** and L, and similarly for $N(a, b)$ and max **AFB**. In sieving we can at best approximate this definition.

6.9. *Remark.* We choose the ordering of (p, r) values for the special-q method as suggested by Pollard [13]: all (p, r) from **AFB** are considered smaller than any $(p, m \bmod p)$ from **RFB**; the p's from **RFB** are placed into their natural order. Then the special q's are chosen from among the rational primes p. This means that the entire algebraic factor base is used at every step. This may not be optimal. It might be better to interleave the rational and algebraic factor bases, with all pairs (p, r) ordered simply by the value of p. We have not explored the practical import of this approach.

6.10. *Remark.* By definition, all special values of (q, s) come from **RFB** or **AFB**. It may happen that $a - bm$ has a large prime factor from outside **RFB**; this factor is not the special q! If we were to include large primes in our ordering of (p, r) values, then any pair with a large p would have to be considered smaller than any pair from **RFB** or **AFB** with respect to this ordering.

6.11. *Remark.* We generate short vectors in a given lattice as in [7, exercise 3.3.4–5]. Except where certain intermediate values are 0 modulo q we accept the resulting vectors as "short enough."

6.12. *Remark.* Initially we suspected that the special-q method would require larger factor bases to break even on the number of reports, even though the total time spent should be smaller. This would make the matrix reduction step slightly more difficult. This was, however, only a consequence of our initial choice of c interval, which was fairly short. The newer version of the program does not require substantially larger factor bases.

6.13. *Remark.* In this section we have taken $T_1, \ldots, T_6$ as positive constant multiples of some unspecified unit of time. In practice this is a reasonable assumption, but in some theoretical machine models these times are not constant. For example, in a bit-oriented model of computation, each T_i is between $\log \max P$ and $\log \max P \log^4 \log \max P$, if we use fast arithmetic for the basic steps. Our conclusion that special-q is asymptotically beneficial holds for any model where all the ratios T_i/T_j grow more slowly than $\log \max P$.

7. TRIAL DIVIDING

We retain the notation of the previous section. The goal of the trial division stage is to produce (a, b) pairs, with a and b relatively prime integers that fit into a computer word, such that b is positive, $a - bm$ is rat-smooth, and $N(a, b) = f(a/b)b^d$ is alg-smooth—prec-smooth pairs, for short. Here $a - bm$ is rat-smooth if all its prime factors are smaller than max RFB, except for at most one large prime factor between max RFB and L. Similarly $N(a, b)$ is alg-smooth if all its prime factors are smaller than max AFB, except for at most one large prime factor between max AFB and L.

Our trial division program **abra** takes as input the reports OUT from the sieving stage. One hopes that many of the pairs (a, b) produced by the sieving stage have $a \equiv b(m \bmod p) \pmod{p}$ for several primes p, so that $a - bm$ is divisible by several primes p and thus has a good chance of being rat-smooth. (Note that $a - bm$ is always divisible by the special q, which is saved in OUT.) Similarly one hopes that $a \equiv br \pmod{p}$ for several algebraic pairs (p, r). Then

$$N(a, b) = f(a/b)b^d \equiv f(r)b^d \equiv 0 \pmod{p}$$

so that $N(a, b)$ is divisible by several primes p and thus has a good chance of being alg-smooth.

Our MasPar trial division program **abra** processes one pair (a, b) at a time, together with the special q value saved from sieving. If a and b are not coprime then (a, b) is thrown out immediately. Otherwise **abra** computes $(a - bm)/q$ and $N(a, b)$. It then invokes a MasPar routine to trial-divide $(a - bm)/q$ by all primes p in the rational factor base. (The prime 2 is not in the factor base for various reasons; it is trial-divided separately.) Unless $a - bm$ is rat-smooth **abra** goes on to the next (a, b) pair. If $a - bm$ factors properly, **abra** trial-divides $N(a, b)$. If $N(a, b)$ is alg-smooth then **abra** saves the prec-smooth pair (a, b) and its prime factorization in a readable file **ABRA**, together with some useful statistics from the sieving stage. Actually, to make disk space easier to manage, we split **ABRA** into a directory of individual files, each with at most 1000 prec-smooth (a, b) pairs.

Once **abra** is done processing (a, b) pairs we remove the output file OUT.

7.1. *Remark.* A prec-smooth pair is usually called an *ff*, *pf*, *fp*, or *pp* relation [9]; it is *pf* or *pp* iff $a - bm$ has a large prime factor, and it is *fp* or *pp* iff $N(a, b)$ has a large prime factor.

8. Constructing the Matrix

Our matrix construction program **newmat** reads the results **ABRA** of the trial division stage, removes duplicates, and produces a single large file **MAT** representing a matrix modulo 2 in binary format. It uses no other data except **QCB**.

The rows of the matrix are indexed by prec-smooth (a, b) pairs. A row contains the following bits:

(1) A bit always equal to 1.

(2) The log base -1 of the sign of $a - bm$. This bit is 1 in all our runs as $a - bm$ is always negative.

(3) For each prime p, $\mathrm{ord}_p(a - bm) \bmod 2$.

(4) For each prime p dividing c_d, $\mathrm{ord}_p N(a, b) \bmod 2$ if p divides b, 0 otherwise.

(5) For each pair (p, r) with $0 \le r < p$ and p prime, $\mathrm{ord}_p N(a, b) \bmod 2$ if p divides $a - br$, 0 otherwise.

(6) Finally, for each (q, s) in the quadratic character base, the log base -1 of the Legendre symbol $\left(\frac{a - bs}{q}\right)$.

Note that these rows are defined in terms of all primes p, not just those appearing in **RFB** and **AFB**. Of course we do not want to store infinitely long rows inside the computer. So instead of one bit for each prime p, we simply store a list of the primes that divide $a - bm$ to odd powers. We allow up to 18 primes in this list. Similarly we store a list of the primes p that divide $N(a, b)$ to odd powers, together with the corresponding values of r. We allow up to 19 pairs (p, r) in this list. We store the #**QCB** $= 100$ Legendre symbols and two extra bits packed into 14 bytes. Together with control information a row takes exactly 256 bytes to store in this format.

Most of the information in a row is easy to compute from **ABRA**. To compute the Legendre symbols we use a fast Jacobi symbol routine due to Peter Montgomery.

For the reasons behind these rows we refer to [3]. In the next steps (next section) of the algorithm we construct a set S of (a, b) such that the corresponding rows add up to 0 (mod 2). In this case we call S a dependency. If S is any set of (a, b) (with #S even) such that $\prod_{(a,b) \in S}(a - bm)$ and the element $\prod_{(a,b) \in S}(a - b\alpha)$ of our number field are both squares then S will, in fact, be a dependency. Conversely, we hope that if S is constructed as a dependency then $\prod_{(a,b) \in S}(a - bm)$ and $\prod_{(a,b) \in S}(a - b\alpha)$ will both be squares. If the quadratic character base is ridiculously large then this will necessarily be true. In practice a quadratic character base of size #**QCB** $= 50$ would probably suffice for all factorizations ever accomplished with the number field sieve in its present form; our choice of #**QCB** $= 100$ is almost certainly excessive.

8.1. *Remark.* Recall from Remark 5.1 that we might occasionally have a bad (q, s) column. For all we know hardware failures might corrupt several columns. It will no longer be true that if $\prod(a - bm)$ and $\prod(a - b\alpha)$ are squares then the corresponding rows of the computed matrix always add up to 0. But if the number of bad quadratic character columns does not exceed the number of excess quadratic character columns, then any computed dependency S will necessarily produce squares. The only bad effect of such errors is that some fraction of the correct dependencies will be ignored during the matrix reduction stage. This justifies the first author's somewhat careless attitude towards the possibility of errors.

9. REDUCING THE MATRIX

The goal of this step is to produce several independent elements of the nullspace of the matrix constructed in the previous section. Note that the matrix is always taken modulo 2. We could have checked the validity of the rows at this point if we were worried about hardware errors.

We find cycles among the partial relations as described in [10]. This reduces the matrix to the set of rows that will be useful in producing dependencies. The useful matrix can be reduced in various ways, such as structured Gaussian elimination [8]. A problem for the future is that structured Gaussian elimination appears to take time cubic in the number of columns of the matrix, although the constant factor is very small. We tried the quadratic method of Wiedemann [15], but were not able to produce a sufficiently fast implementation for our purposes.

Our matrix reduction consists of two programs: **sgauss**, which runs on a workstation, to reduce the large but sparse matrix of ff's and cycles to a smaller dense matrix, and **gauss**, which runs on the MasPar, to find dependencies in the dense matrix. The output of the matrix reduction stage is a file **SOLNS** containing some sets of pairs (a, b) such that the corresponding rows for each set add up to 0 modulo 2.

10. COMPUTING THE SQUARE ROOT

We now have a file **SOLNS** containing one or more sets S of pairs (a, b) such that $\prod_{(a,b)}(a - bm)$ is an integer square and $\prod_{(a,b)}(a - b\alpha)$ is (we hope) a square in the number field. In the final stage of GNFS we convert each set S into a relation $x^2 \equiv y^2 \pmod{n}$. Heuristically at least one half of these pairs (x, y) will give rise to a nontrivial factor $\gcd(x - y, n)$ of n.

Our **cosqrt** program uses the ideas of Couveignes [4]. We prove in the next section that this procedure works. **cosqrt** begins by reading the inert primes ℓ in **IPB**. As we will see in the next section there is a lower bound on the size of **IPB**; we assume that this lower bound is met. Write $P = \prod \ell$. We next compute (P/ℓ) mod ℓ directly for each ℓ, 16384 at a time, on the MasPar. (In this section, when we say "compute X mod M directly," when X is any sort of product, we mean "multiply the factors of the product X, one at a time, modulo M." We do not actually compute X and then reduce it modulo M.) Then we invert P/ℓ modulo ℓ and keep the result, k_ℓ. We also compute P mod n directly for later use.

We read a set of pairs (a, b) that together form a dependency. For each of the inert primes ℓ we compute $\gamma_\ell = g'(\omega)^2 \prod_{(a,b)}(c_d a - b\omega)$ mod ℓ directly inside the number ring $\mathbf{Z}[\omega]/g(\omega)$. We also trial-divide each $a - bm$ and each $N(a, b)$ as in **abra** (see Section 7); we tally up the powers of each prime p dividing $a - bm$

and dividing $N(a, b)$. The tallies are forced to be even, so we end up with

$$\prod_{(a,b)} (a - bm) = \prod_p p^{2r_p} \quad \text{and} \quad \prod_{(a,b)} N(a, b) = \prod_p p^{2a_p}.$$

Next we compute $w = \prod_p p^{r_p} \bmod n$ directly, as well as

$$N_\ell = \text{disc}(g) c_d^{(d-1)\#S/2} \prod_p p^{a_p} \bmod \ell$$

for each inert prime ℓ. We will show in the next section that each N_ℓ is nonzero. Next we set

$$\beta_\ell = \gamma_\ell^{1+\ell(1+\cdots+\ell^{d-2})/2} / N_\ell$$

inside the number ring modulo ℓ; here the division $/N_\ell$ means multiplication by the reciprocal of N_ℓ modulo ℓ.

Write $\beta_\ell = \sum_i b_{\ell,i} \omega^i$. We process each i from 0 through $d - 1$ in turn. For each i we compute an approximation to

$$K_i = \sum_\ell b_{\ell,i} k_\ell / \ell$$

in floating-point arithmetic. Next we compute $Z_i = \text{round}(K_i)$, the integer nearest K_i.

All further computations in **cosqrt** are modulo n. For each i we compute

$$S_i = \sum_\ell b_{\ell,i} k_\ell \frac{P}{\ell} \bmod n;$$

here $(P/\ell) \bmod n$ equals the product of $P \bmod n$ and the multiplicative inverse of $\ell \bmod n$. Finally we compute

$$x = c_d^{d-2+\#S/2} f'(m) w \bmod n$$

and

$$y = \sum_i (S_i - Z_i P)(c_d m)^i \bmod n.$$

We have $x^2 \equiv y^2 \bmod n$, and we finish by checking whether $x - y$ has a factor in common with n. If n is not completely factored, **cosqrt** continues on to the next dependency from **SOLNS**.

10.1. *Remark.* It is the computation of β_ℓ that forces the degree d to be odd. See Section 13 for further comments.

10.2. *Remark.* Some of the ways in which f could conceivably be bad will show up at this stage, though they will rarely affect the factorization. It could turn out that $x \equiv y \equiv 0 \pmod{n}$. In this case, either $f'(m)$, c_d, or one of the primes p in **RFB** will have a factor in common with n.

10.3. *Remark.* If the parameters #QCB and #IPB are not chosen large enough then it may happen that $x^2 \not\equiv y^2 \pmod{n}$. In general (see next section) all values K_i should be extremely close to integers; if they are not then #IPB is too small.

10.4. *Remark.* We can compute N_ℓ without computing disc(g). For

$$\mathrm{disc}(g) \equiv (-1)^{d(d-1)/2} g'(\omega)^{(\ell^d - 1)/(\ell - 1)} \pmod{\ell}$$

if ℓ is any inert prime. This computation must be performed in the number ring, though the result will be an integer. Note that the factor $(-1)^{d(d-1)/2}$ can be omitted, if it is omitted for all ℓ consistently; for reasons explained in the next section, disc(g) need be computed only up to sign.

11. Proof that the square root works

In this section we verify the assertions of the previous section, along the lines of [4]. We retain all notation of the previous section.

We write $N(\delta)$ for the norm of an element δ of our number ring. We will need three properties of the norm: $N(\delta\delta') = N(\delta)N(\delta')$; $N(g'(\omega)) = \pm \mathrm{disc}(g)$; and $N(c_d a - b\omega)$ equals c_d^{d-1} times the norm polynomial $N(a, b)$ defined in Section 3.

Set $\gamma = g'(\omega)^2 \prod_{(a,b)} (c_d a - b\omega)$, so that $\gamma_\ell \equiv \gamma \pmod{\ell}$. If all has gone well γ is in fact a square inside the number ring. Define $\beta = \sum_i b_i \omega^i$ as the unique square root of γ such that $N(\beta)$ has the same sign as $\mathrm{disc}(g)c_d^{(d-1)\#S/2}$; here to ensure uniqueness we need the fact that d is odd. (This is what we call the positive square root method.) Now

$$N(\gamma) = N(g'(\omega))^2 N\Big(\prod_{(a,b)} (c_d a - b\omega)\Big)$$

$$= \mathrm{disc}(g)^2 c_d^{(d-1)\#S} \prod_{(a,b)} N(a, b)$$

$$= \mathrm{disc}(g)^2 c_d^{(d-1)\#S} \prod_p p^{2a_p}.$$

By choice of β we must have

$$N(\beta) = \mathrm{disc}(g)c_d^{(d-1)\#S/2} \prod_p p^{a_p}.$$

Thus $N_\ell \equiv N(\beta) \pmod{\ell}$. It is a pleasant fact of life that

$$N(\beta) \equiv \beta^{(\ell^d - 1)/(\ell - 1)} \pmod{\ell}.$$

We now work modulo ℓ:

$$\beta N_\ell \equiv \beta N(\beta) \equiv \beta \beta^{(\ell^d - 1)/(\ell - 1)} = \gamma^{1 + \ell(1 + \cdots + \ell^{d-2})/2} \equiv \gamma_\ell^{1 + \ell(1 + \cdots + \ell^{d-2})/2} \equiv \beta_\ell N_\ell.$$

We will show later that N_ℓ cannot be divisible by ℓ. Therefore $\beta \equiv \beta_\ell$ (mod ℓ). Of course this means that $b_i \equiv b_{\ell,i}$ (mod ℓ) for each i.

The remaining computations described in the previous section are an application of the Chinese remainder theorem. By construction $k_\ell P/\ell$ is congruent to 1 modulo ℓ and congruent to 0 modulo any $\ell' \neq \ell$. Define

$$R_i = \sum_\ell b_{\ell,i} k_\ell \frac{P}{\ell},$$

so that $S_i = R_i \bmod n$. Since $R_i \equiv b_{\ell,i}$ (mod ℓ) for each ℓ, we must have $R_i \equiv b_i$ (mod P). Define $Z_i' = (R_i - b_i)/P$.

We now show that $Z_i' = Z_i$. Assume that the product $P = \prod \ell$ is large enough that $-P/2 < b_i < P/2$ for each i. We must choose #IPB large enough that this is satisfied. Now round$(b_i/P) = 0$, so round$(R_i/P) = Z_i'$. But by definition of R_i

$$\frac{R_i}{P} = \sum_\ell \frac{b_{\ell,i} k_\ell}{\ell} = K_i.$$

Therefore

$$Z_i' = \text{round}(R_i/P) = \text{round}(K_i) = Z_i$$

as desired.

So $b_i = R_i - Z_i'P = R_i - Z_iP \equiv S_i - Z_iP$ (mod n). We substitute this into the definition of y:

$$y \equiv \sum_i b_i(c_d m)^i \pmod{n}.$$

Define a homomorphism φ from $\mathbf{Z}[\omega]$ to $\mathbf{Z}/n$ by $\varphi(\omega) = c_d m$. Then

$$\varphi(g(\omega)) = g(c_d m) = f(m)c_d^{d-1} \equiv 0 \pmod{n}$$

so φ induces a homomorphism, which we also label φ, from our number ring $\mathbf{Z}[\omega]/g(\omega)$ to $\mathbf{Z}/n$. We compute modulo n:

$$y^2 \equiv \left(\sum_i b_i(c_d m)^i\right)^2 \equiv \left(\varphi\left(\sum_i b_i \omega^i\right)\right)^2$$

$$= \varphi(\beta)^2 = \varphi(\gamma) = \varphi(g'(\omega)^2 \prod_{(a,b)} (c_d a - b\omega))$$

$$= g'(c_d m)^2 \prod_{(a,b)} (c_d a - bc_d m) = (c_d^{d-2} f'(m))^2 c_d^{\#S} \prod_{(a,b)} (a - bm)$$

$$= f'(m)^2 c_d^{\#S+2(d-2)} \prod_p p^{2r_p}$$

$$\equiv x^2.$$

Let us now prove the lemma that N_ℓ is nonzero modulo ℓ. By construction ℓ does not divide disc(g) or c_d. Furthermore, any prime p appearing in $\prod p^{a_p}$ must come from a pair (p, r) in AFB, so that f has a root modulo p. But f is irreducible modulo ℓ. So N_ℓ has no factors divisible by ℓ.

11.1. *Remark.* As we noted above IPB must be large enough that $P = \prod \ell$ is at least twice b_i in absolute value. Certainly the coefficients of γ are comparable to $(ca)^{\#S}$ where c reflects the coefficients of g and a reflects the values of a and b in our relations. Thus the coefficients of β are roughly comparable to $(ca)^{\#S/2}$. We want P to be larger than this. We choose #IPB loosely based on rough estimates of ca and the average value of ℓ. At worst x^2 and y^2 will not be congruent modulo n and we will try a larger inert prime base. Notice that IPB can be chosen smaller if the average ℓ is increased.

12. EXAMPLES

In this section we list some results obtained with our GNFS implementation, and we describe the resources used for these results.

We used GNFS to factor the 145-digit number $(2^{488}+1)/257$. Its prime factors, found at 14:50 EDT on 23 July 1992, are p_{49} and p_{97}, where

$$p_{49} = 1035\,81787\,79260\,14488\,58713\,38184\,91976\,75938\,90347\,64353$$

and

$$p_{97} = 30\,02073\,75742\,87773\,82273\,85792\,23855\,12797\,76379\,27232\,66417$$
$$65602\,50215\,27116\,98977\,99529\,50182\,55653\,75418\,50817.$$

As of August 1992 this was the record non-networked factorization of any difficult number, i.e., any number with no factors under 40 digits.

We proved p_{49} and p_{97} prime both with the Jacobi sum primality test implementation of Bosma and Van der Hulst [2] and with Morain's ECPP implementation of the elliptic-curve primality test [12]. To see if the $p \pm 1$ methods would have worked, we factored $p_{49} \pm 1$ and $p_{97} \pm 1$. It turns out that $p_{49} + 1$ has the large prime factor $(p_{49} + 1)/6$, $p_{49} - 1$ has the large prime factor

$$40\,69106\,49418\,15554\,84740\,76559,$$

$p_{97} + 1$ has the large prime factor $(p_{97} + 1)/474$, and $p_{97} - 1$ has a few large prime factors, including

$$p_{25} = 64969\,31962\,87009\,71159\,63241$$

and

$$p'_{25} = 10512\,71418\,58017\,46720\,49443.$$

Actually it was not entirely trivial to factor $p_{97} - 1$: after a few minutes of the elliptic curve method we were stuck with the 66-digit number

$$346838\,1582437308\,8881132906\,1576183794\,4797477331\,0360327973\,7735819099.$$

To complete this factorization with the elliptic curve method took about three more hours. We also factored the number on the MasPar, with both QS and (without any tuning) GNFS, which was the first truly general GNFS factorization obtained with our implementation; QS succeeded in a few minutes, and GNFS succeeded in a few hours. The prime factors of this 66-digit number are 50781286063727873, p_{25}, and p'_{25}.

At 21:47 EDT on 15 September 1992 we found the prime factors of the 151-digit number $(2^{503} + 1)/3$. They are p_{55} and a p_{97} different from the one above:

$$p_{55} = 20497\,44746\,26356\,86465\,84566\,17590\,83859\,07415\,01232\,90052\,98331$$

and

$$p_{97} = 42\,58599\,33875\,58828\,53705\,02226\,73947\,72292\,11842\,80651\,25200$$
$$53270\,05475\,15369\,56453\,88200\,68351\,21461\,75553\,45713.$$

It is a debatable point whether or not this factorization beats the record set by the factorization of the ninth Fermat number [8]: the composite factor $(2^{512} + 1)/2424833$ of F_9 has only 148 digits, but was just as hard to factor as the 155-digit number $2^{512} + 1$, whereas here we factored a 151-digit number for the price of the 152-digit number $2^{503} + 1$.

We set an unequivocal new factoring record by factoring the 158-digit Mersenne number $2^{523} - 1$. The prime factors, found at 19:30 EDT on 24 October 1992, are p_{69} and p_{90}, where

$$p_{69} = 1601\,88778\,31320\,21186\,10543\,68536\,88786\,88932$$
$$82870\,11365\,01444\,93221\,74680\,39063$$

and

$$p_{90} = 17141\,76918\,61249\,19812\,83170\,96534\,32211\,64761\,65056$$
$$71863\,03450\,94896\,62036\,78600\,06486\,97710\,18595\,04089.$$

We gratefully acknowledge Andrew Odlyzko's help with the sieving and trial division for this record factorization. He did about one third of the work on the MasPar at AT&T Bell Laboratories; the MasPar at Bellcore did the rest.

We proved p_{55}, p_{97}, p_{69}, and p_{90} prime with the Jacobi sum primality test implementation from [2].

In Table 1 our parameter choices and other details concerning these factorizations are reported. We did not include the 66-digit general factorization in the table, as we spent no time tuning it. (We ended up not needing any of the partial relations we generated!)

Our choices for #QCB and #IPB were not tuned at all. Both RFB and IPB use four bytes per prime, and AFB uses eight bytes per (p, r) pair. So for all three numbers these files take at most a few MByte of disk space.

For the first number our MasPar sieving program, **pnfls**, processed over $2 \cdot 10^8$ values of $(q, e, (p, r))$ per second, or over 12,000 values per processing element (PE) per second (where $2 \cdot 10^8 \approx 199947 \cdot 5888 \cdot 66161/(4.5 \cdot 24 \cdot 3600)$). This figure should be taken as somewhat variable because we sieve only up to q on the rational side, and because smaller values of p required more than one hit on the

TABLE 1. *Data on the factorizations*

	$(2^{488}+1)/257$	$(2^{503}+1)/3$	$2^{523}-1$
n			
# digits of n	148	151	158
f	X^5+4	$8X^5+1$	$8X^5-1$
m	2^{98}	2^{100}	2^{104}
max RFB	1300000	1300000	1600000
#RFB	100020	100020	121126
max AFB	1300000	1300000	1600000
#AFB	99827	99827	120909
#QCB	100	100	100
#IPB	163909	163909	196586
L	10^8	10^8	10^8
c interval	$[-5000, 5000)$	$[-5000, 5000)$	$[-14000, 14000)$
$\{e\}$	$[1, 5888]$	$[1, 7936]$	$[1, 12800]$
#Q	66161	82037	≈ 65000
min Q	≈ 400000	≈ 200000	≈ 700000
run time pnfls	4.5 days	6.5 days	14 days
space pnfls	24 KByte per PE	24 KByte per PE	32 KByte per PE
#OUT	$18 \cdot 10^6$	$55 \cdot 10^6$	$13 \cdot 10^7$
run time abra	1.7 days	3 days	7 days
space abra	4 KByte per PE	4 KByte per PE	4 KByte per PE
# ff's	60253	51905	not kept
# ffq's	32060	40999	not kept
# pf's	367402	348074	not kept
# pfq's	20573	59192	not kept
# fp's	324296	289665	not kept
# fpq's	141171	213419	not kept
# pp's	1681923	1920259	not kept
# ppq's	2057	239124	not kept
size ABRA	721 MByte	867 MByte	not kept
# distinct ff's	65626	57728	> 70000
# distinct partials	2409258	2713637	$\approx 3.5 \cdot 10^6$
size MAT	624 MByte	709 MByte	895 MByte
total # cycles	142657	149813	> 210000
sparse matrix	208283×199947	207541×199947	280000×242135
(run time sgauss)	(16 hours)	(16 hours)	(2 days)
dense matrix	not kept	74100×73900	94000×93900
size dense matrix	not kept	685 MByte	1.1 GByte
run time gauss	not kept	2.1 hours	5 hours
#S	not counted	≈ 257000	≈ 283000
run time cosqrt	7.5 hours	7.5 hours	9 hours
# cosqrt trials	3	5	3
total run time	7.5 days	12 days	22 days

c interval from (6.7); we used primes as low as 23. In any case, as each processor performs about $2 \cdot 10^5$ additions per second, it appears that we did the work of about 16 additions per $(q, e, (p, r))$. For the second number the sieving speed was comparable. The version of **pnfls** that we used for the first two numbers took 2 bytes per c. At some expense in run time we reduced this to 1 byte in the version of **pnfls** that we used for the third number. With slightly more space per PE, this allowed us to almost triple the c interval. As a result the sieving speed for the third number was somewhat slower, but the yield was much higher.

OUT, the binary output from sieving, uses 24 bytes per report. With well-tuned cutoff parameters there are not too many more reports than actual relations. For the first number we chose the cutoffs so as to eliminate many, perhaps most, of the pp relations. For the other two numbers the cutoffs were chosen more conservatively. This led to far fewer relations per report, but to more partials and cycles. For the last number this could have led to storage problems for **OUT**. We avoided this by sieving in batches of special q values, removing **OUT** after running **abra** for each batch, so that **OUT** never used more than eighty MByte of disk space at a time.

For the first number **abra** processed 7500 reports per minute. A newer version of **abra**, which was used for the other two numbers, achieved almost twice that speed. Of course, many of the reports were cut out after trial division by the primes of **RFB** and did not require division by the primes of **AFB**.

ABRA, the readable relations file resulting from **abra**, takes a lot of disk space. The "q" suffix of the ff's, pf's, fp's, and pp's (see Remark 7.1) means that the factorization of $a - bm$ ended up with a larger prime than q from **RFB**. In principle such relations have the "wrong" q and are not useful, as they should also be found with the "right" q. However, our sieving bounds were so high that many relations are not actually hit during sieving by their special q's; and our sieving bounds were so low that many ff's and fp's were caught by a lower q than their special q's. The fact that our sieving bounds were simultaneously so high and so low stems from the inherent inaccuracy of the logarithmic approximations we used; see Section 6. Another explanation for the occurrence of relations with the "wrong" q is that we searched somewhat different areas of the (a, b) plane for different q.

Because of the occurrence of ffq's, pfq's, fpq's, and ppq's, duplicate relations were found, which were removed by **newmat** before it constructed the quadratic characters. The run time of **newmat** was reasonably short, and is not given in Table 1. By far the most costly activity in this stage was I/O. The run time for **sgauss** is parenthesized because this was the only substantial step that was not carried out on a MasPar but on an ordinary workstation. We note that the dense matrix for the second number is larger than the dense matrix involved in factorization of the ninth Fermat number [8]. Its reduction produced considerably more than $74100 - 73900$ dependencies. For the last number the dense matrix did not fit in the MasPar memory (which is only 1 GByte), so we had to upgrade our in-core matrix eliminator to a more flexible but slower program that stores intermediate results on disk. Again we found many more dependencies than the approximately $94000 - 93900$ that we expected. This might be due to our very conservative choice of #QCB.

Our cosqrt is not inherently faster than the simple method stated in [3] but it is much more amenable to parallelization. The cosqrt run time in the Table refers to the processing time per dependency. Most of this time was used to compute γ_ℓ.

We also factored a 123-digit composite factor of $2^{491} + 1$, as well as $(11^{131} - 1)/2630$. Of course all these numbers could have been handled by SNFS, even though $\mathbf{Q}(\sqrt[5]{11})$, the field of choice for $(11^{131} - 1)/2630$, has class number 5, as reported by R. D. Silverman.

13. FUTURE DIRECTIONS

The crucial problem in GNFS is finding a good polynomial. We cannot overestimate the practical importance of searching for a polynomial a few digits better than random. According to theoretical estimates [3] our polynomials are nowhere near optimal. As polynomials improve, GNFS will become more and more competitive with QS; a much better polynomial-finding method could make GNFS the leader for general 100-digit numbers.

As indicated in Section 4, without m_2 from [3] we can in theory reduce the coefficients by several digits. With m_2 we can in theory reduce the coefficients by twice as many digits. So we should incorporate m_2 into our GNFS implementation as polynomial-searching methods improve.

The restriction that d must be odd is not at all helpful, as $d = 4$ appears to be a very good choice for numbers around 100 digits. If d is even, is there a way to uniquely specify β, analogous to the requirement that $N(\beta)$ have a certain sign, that can be tested easily modulo any ℓ? Or is there a fast way to determine the two values of β mod $\ell\ell'$? If either of these problems is solved then we can use even degrees. Of course, the method in [3] is reasonably fast, given a fast multidigit multiplication routine; but it does not seem to parallelize easily.

Don Coppersmith has suggested a method that will work for even degrees at the expense of some arithmetic on huge numbers (much less than in [3] for $d > 1$). We illustrate it for $d = 4$ under the assumption that (say) b_1 is not divisible by n. Compute γ in the number ring modulo each ℓ. By standard methods compute some square root β_ℓ of γ in each of the finite fields $(\mathbf{Z}/\ell\mathbf{Z})[\omega]/g(\omega)$. The square roots need not be consistent over different ℓ, but the products $b_{\ell,j} b_{\ell,1}$ for $j = 0, 1, 2, 3$ are consistent. So we combine these products for $j \neq 1$ into values $b_j b_1$ modulo n as in Section 11. Using the standard Chinese remainder theorem, with arithmetic on huge numbers, we combine the squares $b_{\ell,1}^2$ into a value b_1^2. We then choose one of the two square roots b_1 and have enough information to put together a consistent b_j mod n for $j = 0, 1, 2, 3$ as desired. It remains to be seen how much work this method will require in practice.

We have not yet incorporated certain practical improvements. There are a significant number of "free relations" [9]; we should use them. We judge a special q as productive if it is not in the bottom fraction of RFB, as per [13]; we should pay more attention to the reduced lattice vectors. We should experiment with interleaving the factor bases as suggested in Remark 6.9. We should compute the actual number of quadratic characters needed in various real examples, then use only a few more than the maximum, rather than 100. We should use (p, ∞) pairs in the sieving when possible.

What can be said about the sizes of the coefficients b_i? It would be nice to have accurate estimates—taking the distribution of (a, b) into account—so that we could select #IPB sensibly. We have certainly wasted time on the MasPar processing excessively many values of ℓ.

Our parameters are not chosen optimally. In particular the algebraic factor base should probably be much larger than the rational factor base for best results. It would help greatly if we could quickly and reliably estimate, to within a few percent, the number of relations that will be produced by the sieving, given POLY and all relevant parameters.

It is tempting to try to construct pairs (a, b) with a high likelihood of being prec-smooth. For instance, if a has a large factor in common with m, then $a - bm$ will also have that factor (this is equivalent to dividing b by the factor, as suggested by R. D. Silverman). Unfortunately it appears that $N(a, b)$ becomes too big. Similarly if a/b is extremely close to a root of f then $N(a, b)$ will be small, but then $a - bm$ will not remain small enough. It is not clear whether considerations like these are the path to an improved number field sieve or merely minor curiosities.

REFERENCES

1. L. M. Adleman, *Factoring numbers using singular integers*, Proc. 23rd Annual ACM Symp. on Theory of Computing (STOC), New Orleans, May 6–8, 1991, 64–71.
2. W. Bosma, M.-P. van der Hulst, *Primality proving with cyclotomy*, Universiteit van Amsterdam, 1990.
3. J. P. Buhler, H. W. Lenstra, Jr., C. Pomerance, *Factoring integers with the number field sieve*, this volume, pp. 50–94.
4. J.-M. Couveignes, *Computing a square root for the number field sieve*, this volume, pp. 95–102.
5. J. A. Davis, D. B. Holdridge, *Factorization using the quadratic sieve algorithm*, Tech. Report SAND 83-1346, Sandia National Laboratories, Albuquerque, New Mexico, 1983.
6. B. Dixon, A. K. Lenstra, *Factoring integers using SIMD sieves*, Advances in Cryptology, Eurocrypt '93, to appear.
7. D. E. Knuth, *The art of computer programming*, volume 2, *Seminumerical algorithms*, second edition, Addison-Wesley, Reading, Massachusetts, 1981.
8. A. K. Lenstra, H. W. Lenstra, Jr., M. S. Manasse, J. M. Pollard, *The factorization of the ninth Fermat number*, Math. Comp. **61** (1993), to appear.
9. A. K. Lenstra, H. W. Lenstra, Jr., M. S. Manasse, J. M. Pollard, *The number field sieve*, this volume, pp. 11–42.
10. A. K. Lenstra, M. S. Manasse, *Factoring with two large primes*, Math. Comp., to appear.
11. H. W. Lenstra, Jr., *Factoring integers with elliptic curves*, Ann. of Math. **126** (1987), 649–673.
12. F. Morain, *Implementation of the Goldwasser-Kilian-Atkin primality testing algorithm*, INRIA report 911, INRIA-Rocquencourt, 1988.
13. J. M. Pollard, *The lattice sieve*, this volume, pp. 43–49.
14. C. Pomerance, *The quadratic sieve factoring algorithm*, Lecture Notes in Comput. Sci. **209** (1985), 169–182.

15. D. Wiedemann, *Solving sparse linear equations over finite fields*, IEEE Trans. Inform. Theory **32** (1986), 54–62.

5 BREWSTER LANE, BELLPORT, NY 11713, U.S.A.
E-mail address: brnstnd@nyu.edu

ROOM MRE-2Q334, BELLCORE, 445 SOUTH STREET, MORRISTOWN, NJ 07960, U.S.A.
E-mail address: lenstra@bellcore.com

THE ILLUSTRATION ON THE FRONT COVER

The illustration on the front cover is inspired by the Remark on page 53. It depicts the spectrum of the subring of $\mathbf{Z} \times \mathbf{Z}[\alpha]$ generated by (m, α). In the informal explanation below we let *ring* stand for *commutative ring with 1*.

In contemporary algebraic geometry one associates with every ring a topological space, called the *spectrum* of the ring. This is done in such a way that the elements of the ring can be thought of as functions defined on that space. This makes it possible to use geometric intuition when dealing with algebraic objects. This is useful not only in algebraic geometry, but also in number theory, though not yet in integer factoring.

As an example, consider the ring $\mathbf{C}[X, Y]$ of polynomials in two variables X, Y over the field $\mathbf{C}$ of complex numbers. One can view elements of this ring as polynomial functions defined on $\mathbf{C} \times \mathbf{C}$, with values in $\mathbf{C}$; and the spectrum of $\mathbf{C}[X, Y]$ is, accordingly, almost the same as $\mathbf{C} \times \mathbf{C}$.

As another example, take the ring $\mathbf{Z}$ of ordinary integers. In this case it is less clear how the elements of the ring can be viewed as functions on some space, but algebraic geometry tells us what to do: the space is just the set of prime numbers, and each integer n is thought of as the function that maps a prime number p to the residue class $(n \bmod p)$. Thus, the spectrum of $\mathbf{Z}$ is essentially the same as the set of prime numbers. In general, the spectrum of a ring is equal to the set of prime ideals of the ring. For instance, if n is a positive integer, then the number of points in the spectrum of $\mathbf{Z}/n\mathbf{Z}$ is equal to the number of distinct prime factors of n. For the illustration, we have assumed that n is the product of two different prime numbers, corresponding to the two dots that form the intersection points of the two curves.

The spectrum of a ring can be made into a topological space. This suggests a pictorial representation of rings that is even less precise than pictures of topological spaces in general. Nevertheless, several properties of a ring can be reflected in a picture of its spectrum. For example, the rings $\mathbf{Z}$ and $\mathbf{Z}[\alpha]$ have the property that all of their non-zero prime ideals are maximal ideals. One expresses this by saying that the rings are *one-dimensional*, and their spectra are accordingly represented as curves, which are one-dimensional objects. The smoothness of the curve that goes with $\mathbf{Z}$ expresses that $\mathbf{Z}$ equals the full ring of algebraic integers in the field of rational numbers. The cusp and the node in the curve that represents $\mathbf{Z}[\alpha]$ indicate the possibility of various singularities when the ring is not the full ring of integers in the number field.

The "small" prime ideals of $\mathbf{Z}$ and $\mathbf{Z}[\alpha]$, which play an important role in the number field sieve, are found near the "beginning" of the respective curves.

The spectrum of the product ring $\mathbf{Z} \times \mathbf{Z}[\alpha]$ would just be the disjoint union of the spectra of $\mathbf{Z}$ and $\mathbf{Z}[\alpha]$: two separate curves without intersection points. Restricting to the subring generated by (m, α), which has index n in the full ring, comes down to gluing the two curves together in two points. Those two points correspond to the two prime divisors of n.

To factor n one needs to separate the two points of the spectrum of $\mathbf{Z}/n\mathbf{Z}$ from each other. In terms of the picture, the number field sieve attempts to achieve this by pulling the two curves apart at one intersection point while keeping them joined at the other.

The main technical tool of the number field sieve is a two-dimensional sieving process. This is represented in the background of the picture, which shows two two-dimensional lattices and their intersection.

An illustrated introduction to spectra of rings can be found in *Schemes: the language of modern algebraic geometry* by D. Eisenbud and J. Harris (Brooks/Cole, 1992).

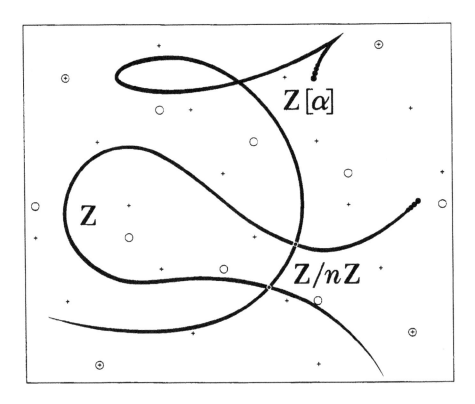

INDEX

Printing: Weihert-Druck GmbH, Darmstadt
Binding: Buchbinderei Schäffer, Grünstadt

Lecture Notes in Mathematics

For information about Vols. 1–1370
please contact your bookseller or Springer-Verlag

Vol. 1412: V.V. Kalashnikov, V.M. Zolotarev (Eds.), Stability Problems for Stochastic Models. Proceedings, 1987. X, 380 pages. 1989.

Vol. 1413: S. Wright, Uniqueness of the Injective III₁Factor. III, 108 pages. 1989.

Vol. 1414: E. Ramirez de Arellano (Ed.), Algebraic Geometry and Complex Analysis. Proceedings, 1987. VI, 180 pages. 1989.

Vol. 1415: M. Langevin, M. Waldschmidt (Eds.), Cinquante Ans de Polynômes. Fifty Years of Polynomials. Proceedings, 1988. IX, 235 pages.1990.

Vol. 1416: C. Albert (Ed.), Géométrie Symplectique et Mécanique. Proceedings, 1988. V, 289 pages. 1990.

Vol. 1417: A.J. Sommese, A. Biancofiore, E.L. Livorni (Eds.), Algebraic Geometry. Proceedings, 1988. V, 320 pages. 1990.

Vol. 1418: M. Mimura (Ed.), Homotopy Theory and Related Topics. Proceedings, 1988. V, 241 pages. 1990.

Vol. 1419: P.S. Bullen, P.Y. Lee, J.L. Mawhin, P. Muldowney, W.F. Pfeffer (Eds.), New Integrals. Proceedings, 1988. V, 202 pages. 1990.

Vol. 1420: M. Galbiati, A. Tognoli (Eds.), Real Analytic Geometry. Proceedings, 1988. IV, 366 pages. 1990.

Vol. 1421: H.A. Biagioni, A Nonlinear Theory of Generalized Functions, XII, 214 pages. 1990.

Vol. 1422: V. Villani (Ed.), Complex Geometry and Analysis. Proceedings, 1988. V, 109 pages. 1990.

Vol. 1423: S.O. Kochman, Stable Homotopy Groups of Spheres: A Computer-Assisted Approach. VIII, 330 pages. 1990.

Vol. 1424: F.E. Burstall, J.H. Rawnsley, Twistor Theory for Riemannian Symmetric Spaces. III, 112 pages. 1990.

Vol. 1425: R.A. Piccinini (Ed.), Groups of Self-Equivalences and Related Topics. Proceedings, 1988. V, 214 pages. 1990.

Vol. 1426: J. Azéma, P.A. Meyer, M. Yor (Eds.), Séminaire de Probabilités XXIV, 1988/89. V, 490 pages. 1990.

Vol. 1427: A. Ancona, D. Geman, N. Ikeda, École d'Eté de Probabilités de Saint Flour XVIII, 1988. Ed.: P.L. Hennequin. VII, 330 pages. 1990.

Vol. 1428: K. Erdmann, Blocks of Tame Representation Type and Related Algebras. XV. 312 pages. 1990.

Vol. 1429: S. Homer, A. Nerode, R.A. Platek, G.E. Sacks, A. Scedrov, Logic and Computer Science. Seminar, 1988. Editor: P. Odifreddi. V, 162 pages. 1990.

Vol. 1430: W. Bruns, A. Simis (Eds.), Commutative Algebra. Proceedings. 1988. V, 160 pages. 1990.

Vol. 1431: J.G. Heywood, K. Masuda, R. Rautmann, V.A. Solonnikov (Eds.), The Navier-Stokes Equations – Theory and Numerical Methods. Proceedings, 1988. VII, 238 pages. 1990.

Vol. 1432: K. Ambos-Spies, G.H. Müller, G.E. Sacks (Eds.), Recursion Theory Week. Proceedings, 1989. VI, 393 pages. 1990.

Vol. 1433: S. Lang, W. Cherry, Topics in Nevanlinna Theory. II, 174 pages.1990.

Vol. 1434: K. Nagasaka, E. Fouvry (Eds.), Analytic Number Theory. Proceedings, 1988. VI, 218 pages. 1990.

Vol. 1435: St. Ruscheweyh, E.B. Saff, L.C. Salinas, R.S. Varga (Eds.), Computational Methods and Function Theory. Proceedings, 1989. VI, 211 pages. 1990.

Vol. 1436: S. Xambó-Descamps (Ed.), Enumerative Geometry. Proceedings, 1987. V, 303 pages. 1990.

Vol. 1437: H. Inassaridze (Ed.), K-theory and Homological Algebra. Seminar, 1987–88. V, 313 pages. 1990.

Vol. 1438: P.G. Lemarié (Ed.) Les Ondelettes en 1989. Seminar. IV, 212 pages. 1990.

Vol. 1439: E. Bujalance, J.J. Etayo, J.M. Gamboa, G. Gromadzki. Automorphism Groups of Compact Bordered Klein Surfaces: A Combinatorial Approach. XIII, 201 pages. 1990.

Vol. 1440: P. Latiolais (Ed.), Topology and Combinatorial Groups Theory. Seminar, 1985–1988. VI, 207 pages. 1990.

Vol. 1441: M. Coornaert, T. Delzant, A. Papadopoulos. Géométrie et théorie des groupes. X, 165 pages. 1990.

Vol. 1442: L. Accardi, M. von Waldenfels (Eds.), Quantum Probability and Applications V. Proceedings, 1988. VI, 413 pages. 1990.

Vol. 1443: K.H. Dovermann, R. Schultz, Equivariant Surgery Theories and Their Periodicity Properties. VI, 227 pages. 1990.

Vol. 1444: H. Korezlioglu, A.S. Ustunel (Eds.), Stochastic Analysis and Related Topics VI. Proceedings, 1988. V, 268 pages. 1990.

Vol. 1445: F. Schulz, Regularity Theory for Quasilinear Elliptic Systems and – Monge Ampère Equations in Two Dimensions. XV, 123 pages. 1990.

Vol. 1446: Methods of Nonconvex Analysis. Seminar, 1989. Editor: A. Cellina. V, 206 pages. 1990.

Vol. 1447: J.-G. Labesse, J. Schwermer (Eds), Cohomology of Arithmetic Groups and Automorphic Forms. Proceedings, 1989. V, 358 pages. 1990.

Vol. 1448: S.K. Jain, S.R. López-Permouth (Eds.), Non-Commutative Ring Theory. Proceedings, 1989. V, 166 pages. 1990.

Vol. 1449: W. Odyniec, G. Lewicki, Minimal Projections in Banach Spaces. VIII, 168 pages. 1990.

Vol. 1450: H. Fujita, T. Ikebe, S.T. Kuroda (Eds.), Functional-Analytic Methods for Partial Differential Equations. Proceedings, 1989. VII, 252 pages. 1990.

Vol. 1451: L. Alvarez-Gaumé, E. Arbarello, C. De Concini, N.J. Hitchin, Global Geometry and Mathematical Physics. Montecatini Terme 1988. Seminar. Editors: M. Francaviglia, F. Gherardelli. IX, 197 pages. 1990.

Vol. 1452: E. Hlawka, R.F. Tichy (Eds.), Number-Theoretic Analysis. Seminar, 1988–89. V, 220 pages. 1990.

Vol. 1453: Yu.G. Borisovich, Yu.E. Gliklikh (Eds.), Global Analysis – Studies and Applications IV. V, 320 pages. 1990.

Vol. 1454: F. Baldassari, S. Bosch, B. Dwork (Eds.), p-adic Analysis. Proceedings, 1989. V, 382 pages. 1990.

Vol. 1455: J.-P. Françoise, R. Roussarie (Eds.), Bifurcations of Planar Vector Fields. Proceedings, 1989. VI, 396 pages. 1990.

Vol. 1456: L.G. Kovács (Ed.), Groups – Canberra 1989. Proceedings. XII, 198 pages. 1990.

Vol. 1457: O. Axelsson, L.Yu. Kolotilina (Eds.), Preconditioned Conjugate Gradient Methods. Proceedings, 1989. V, 196 pages. 1990.

Vol. 1458: R. Schaaf, Global Solution Branches of Two Point Boundary Value Problems. XIX, 141 pages. 1990.

Vol. 1459: D. Tiba, Optimal Control of Nonsmooth Distributed Parameter Systems. VII, 159 pages. 1990.